PRODUCTION AND NEUTRALIZATION OF NEGATIVE IONS AND BEAMS

Related Titles from AIP Conference Proceedings and the Subseries on Accelerators, Beams, and Instrumentation

592 High Quality Beams: Joint US-CERN-JAPAN-RUSSIA Accelerator School
Edited by S. I. Kurokawa, S. Y. Lee, J. Miles, E. A. Perevedentsev, November 2001,
0-7354-0034-2

572 Electron Beam Ion Sources and Traps and Their Applications:
8th International Symposium
Edited by Krsto Prelec, June 2001, 0-7354-0011-3

480 Space Charge Dominated Beam Physics for Heavy Ion Fusion
Edited by Yuri K. Batygin, June 1999, 1-56396-860-6

473 Heavy Ion Accelerator Technology: Eighth International Conference
Edited by Kenneth W. Shepard, April 1999, 1-56396-806-1

439 Production and Neutralization of Negative Ions and Beams:
Eighth International Symposium/Production and Application of Light Negative Ions:
Seventh European Workshop
Edited by Claude Jacquot, August 1998, 1-56396-737-5

To learn more about these titles, or the AIP Conference Proceedings Series, please visit the webpage **http://proceedings.aip.org/proceedings**

PRODUCTION AND NEUTRALIZATION OF NEGATIVE IONS AND BEAMS

Ninth International Symposium on the Production and Neutralization of Negative Ions and Beams

Gif-sur-Yvette, France   30–31 May 2002

EDITOR
Martin P. Stockli
*Spallation Neutron Source
Oak Ridge, Tennessee*

SPONSORING ORGANIZATIONS
Commissariat à l'Energie Atomique, Saclay, France
Spallation Neutron Source, Oak Ridge National Laboratory, USA
U.S. Department of Energy

Melville, New York, 2002
AIP CONFERENCE PROCEEDINGS ■ VOLUME 639

Editor:

Martin P. Stockli
Spallation Neutron Source
115 Union Valley Road
Oak Ridge, TN 37830
USA

E-mail: stockli@sns.gov

The articles on pp. 47-60 and 160-174 were authored by U.S. Government employees and are not covered by the below mentioned copyright.

Authorization to photocopy items for internal or personal use, beyond the free copying permitted under the 1978 U.S. Copyright Law (see statement below), is granted by the American Institute of Physics for users registered with the Copyright Clearance Center (CCC) Transactional Reporting Service, provided that the base fee of $19.00 per copy is paid directly to CCC, 222 Rosewood Drive, Danvers, MA 01923. For those organizations that have been granted a photocopy license by CCC, a separate system of payment has been arranged. The fee code for users of the Transactional Reporting Service is: 0-7354-0094-6/02/$19.00.

© 2002 American Institute of Physics

Individual readers of this volume and nonprofit libraries, acting for them, are permitted to make fair use of the material in it, such as copying an article for use in teaching or research. Permission is granted to quote from this volume in scientific work with the customary acknowledgment of the source. To reprint a figure, table, or other excerpt requires the consent of one of the original authors and notification to AIP. Republication or systematic or multiple reproduction of any material in this volume is permitted only under license from AIP. Address inquiries to Office of Rights and Permissions, Suite 1NO1, 2 Huntington Quadrangle, Melville, N.Y. 11747-4502; phone: 516-576-2268; fax: 516-576-2450; e-mail: rights@aip.org.

L.C. Catalog Card No. 2002112981
ISBN 0-7354-0094-6
ISSN 0094-243X
Printed in the United States of America

CONTENTS

Preface vii
Symposium Organization ix

FUNDAMENTAL PROCESSES

**Electron-Impact Cross Sections for Processes Involving Vibrationally
Excited Diatomic Hydrogen Molecules** 3
    R. Celiberto and A. Laricchiuta
**Influence of Wall Material on VUV Emission from Hydrogen Plasma
in H$^-$ Source** 13
    M. Bacal, M. Glass-Maujean, A. A. Ivanov Jr., M. Nishiura, M. Sasao, and
    M. Wada
**Measurements of H$^-$ Density and Work Function of a Plasma
Electrode in a Negative Ion Source** 21
    M. Nishiura, M. Sasao, and M. Wada
**Study of Isotope Effect on H$^-$/D$^-$ Volume Production in
Low-Pressure H$_2$/D$_2$ Plasmas Using VUV Emission** 28
    O. Fukumasa, Y. Tauchi, Y. Yabuki, S. Mori, and Y. Takeiri

SOURCES

**A Review of Recent H$^-$ Ion Source Work at Rutherford
Appleton Laboratory** 37
    J. W. G. Thomason
**The New HERA H$^-$ RF Volume Source and Selected Results of
Plasma Investigations** 42
    J. Peters
**Design, Operational Experiences, and Beam Results Obtained with
the SNS H$^-$ Ion Source and LEBT at Berkeley Lab** 47
    R. Keller, R. W. Thomae, M. P. Stockli, and R. F. Welton
**Status of an RF Negative Hydrogen Ion Source Using Transformer
Coupled Plasma Source** 61
    I. S. Hong, H. D. Jung, Y. S. Park, and Y. S. Hwang
A High-Intensity H$^-$ Ion Source 67
    K. Volk, H. Klein, A. Maaser, and U. Ratzinger

BEAM FORMATION AND ACCELERATION

**Modeling of Negative Ion Transport in Hydrogen Ion Sources—
Estimation of Extracted H$^-$ Current** 75
    O. Fukumasa, T. Fujioka, and T. Fukuchi

An Inverted Plasma Sheath for the Simulation of the Extraction of
Volume Produced H⁻ 82
    R. Becker
Studies on the Extraction Region of the Type VI RF Driven
H⁻ Ion Source... 90
    P. McNeely, M. Bandyopadhyay, P. Franzen, B. Heinemann, C. Hu,
    W. Kraus, R. Riedl, E. Speth, and R. Wilhelm

DIAGNOSTICS

Studies on a Magnetron Source 115
    D. P. Moehs
Status Report of the Frankfurt H⁻-Test LEBT Including a
Nondestructive Emittance Measurement Device 121
    C. Gabor, A. Jakob, O. Meusel, J. Schäfer, A. Klomp, F. Santić,
    J. Pozimski, H. Klein, and U. Ratzinger
Diagnostics at the Frankfurt H⁻-LEBT 128
    A. Jakob, C. Gabor, O. Meusel, J. Pozimski, H. Klein, and U. Ratzinger
Accurate Estimation of the RMS Emittance from Single Current
Amplifier Data.. 135
    M. P. Stockli, R. F. Welton, R. Keller, A. P. Letchford, R. W. Thomae, and
    J. W. G. Thomason
Emittance Characteristics of High-Brightness H⁻ Ion Sources 160
    R. F. Welton, M. P. Stockli, R. Keller, R. W. Thomae, J. W. G. Thomason,
    J. D. Sherman, and J. Alessi

NEW CONCEPTS

The CEA/Saclay 2.45 GHz Microwave Ion Source for
H⁻ Ion Production 177
    R. Gobin, K. Benmeziane, O. Delferrière, R. Ferdinand, F. Harrault, and
    J. D. Sherman
First Simulations of the Cadarache SINGAP Experiments 184
    H. P. L. de Esch, D. Boilson, R. S. Hemsworth, P. Massmann, and
    L. Svensson
Negative Ion Production by fs, High-Intensity Laser Beam
Interactions with Clusters (*abstract*) 197
    S. D. Moustaizis, P. Balcou, J.-P. Chambaret, D. Hulin, G. Grillon,
    J.-P. Rousseau, and M. Schmidt
Summary Comments 198
    J. D. Sherman

Symposium Program 201
List of Participants.. 203
Author Index 207

Preface

The 9th International Symposium on the Production and Neutralization of Negative Ions and Beams was held at CEA Saclay, close to Paris, France. After a five-year break, this two-day workshop (May 30–31, 2002) continued the series of conferences organized in the United States or western Europe since 1977. The last meeting was held at Giens, France, in 1997 and grouped the 8th International Symposium and the 7th European Workshop on the production and application of light negative ions. At this meeting, it was decided to continue this kind of joint meeting. Thanks go to M. Stockli from SNS, who initiated an open discussion at the last ICIS Conference to continue the series. The June date and the Saclay site were chosen to coincide with EPAC 2002 in Paris, France, in early June.

Spallation source projects (SNS in the United States, ESS in Europe, Joint Project in Japan, and others), based on high-power accelerators, push to develop high brightness negative light ion sources. For this application, the use of negative hydrogen ions is imposed by the compressor ring injection optimization. Currently, these projects are very demanding of ion source developments. This and EPAC probably explain the high participation of the accelerator source community in the Saclay workshop. The fusion community, which organized an international meeting last April, was not as highly represented. However, the developments achieved for many years or currently in progress by these groups are useful for the whole negative ion source community.

A better accelerator and fusion community ratio is expected for future meetings. A good way to improve this ratio would be to organize the next conference as a satellite workshop of the future ICIS Conference in 2003.

We would like to thank the Commissariat à l'Energie Atomique for ensuring the success of this meeting and the members of the International Program Committee who provided valuable advice. We would also especially like to thank Catherine Desailly-Guyard, Anne-Marie Gauriot, and the local organizing committee, who made every effort to guarantee the success of this workshop.

To conclude, we sincerely congratulate all the participants who ensured valuable presentations and fruitful discussions. Finally, special thanks to M. Stockli, who was in charge of the relations with the publisher, and J. D. Sherman for his concluding remarks.

Raphaël GOBIN
CEA/Saclay
DSM/DAPNIA
91 191 Gif sur Yvette
France

Symposium Organization

CHAIR

R. Gobin, Chair   CEA/Saclay (France)
R. Becker, Co-chair   University of Frankfurt (Germany)

INTERNATIONAL PROGRAM COMMITTEE

M. Bacal   Ecole Polytechnique (France)
R. Becker   University of Frankfurt (Germany)
Y. Belchenko   Budker Institute (Russia)
B. Ellingboe   Dublin University (Ireland)
R. Gobin   CEA/Saclay (France)
R. Hemsworth   CEA/Cadarache (France)
T. Inoue   JAERI (Japan)
K.-N. Leung   LBNL (USA)
J. Sherman   LANL (USA)
M. Stockli   SNS (USA)
J. Peters   DESY (Germany)
J. Thomason   RAL (UK)

SYMPOSIUM SECRETARIAT

C. Desailly   DSM/DAPNIA/SACM/CEA/Saclay (France)
A.M. Gauriot   DSM/DAPNIA/SACM/CEA/Saclay (France)

SYMPOSIUM LOCATION

CEA/Saclay/Orme des Merisiers
Orme des Merisiers / Amphithéâtre Claude Bloch
91191 Gif-sur-Yvette Cédex – France

SPONSORS

Commissariat à l'Energie Atomique, Saclay (France)
Spallation Neutron Source, Oak Ridge National Laboratory (USA)
U.S. Department of Energy

FUNDAMENTAL PROCESSES

Electron-Impact Cross Sections for Processes Involving Vibrationally Excited Diatomic Hydrogen Molecules

R. Celiberto* and A. Laricchiuta[†]

Dipartimento di Ingegneria Civile ed Ambientale, Politecnico di Bari, Italy
[†]*IMIP, Sezione Territoriale di Bari, CNR, Italy*

Abstract. Electron-impact cross sections for processes involving vibrationally excited molecules of hydrogen and its isotopes are reviewed, briefly discussing their role in negative ion plasma sources.

1. INTRODUCTION

Dissociative attachment is generally accepted as the main process leading to the production of negative ions H^- (or D^-) in volume plasma sources [1]. This process occurs by the capture of slow electrons (~ 1 eV) by vibrationally excited diatomic hydrogen molecules, with the subsequent formation of a resonant state H_2^-, which can evolve in dissociation yielding finally to a ground state hydrogen atom and a negative ion H^-. The whole process can be sketched as

$$H_2(X^1\Sigma_g^+, v_i) + e \rightarrow H_2^- \rightarrow H(1s) + H^- \tag{1}$$

where the hydrogen molecule is initially in its ground electronic state and in a given vibrational level v_i.

The cross sections for this process, calculated years ago by Bardsley and Wadehra for both H_2 and D_2 [2], present a strong dependence on the vibrational state of the molecule (figure 1), with an increment of several orders of magnitude for high-lying vibrational levels. As a consequence, a large concentration in the plasma of vibrationally excited molecules becomes the fundamental prerequisite for an effective production of negative ions.

Vibrational kinetics thus plays an important role in these systems and the theoretical models of the H^- sources, aimed at the optimization of the experimental conditions, must include all those processes that may affect the vibrational population. In addition, due to the strong non-equilibrium conditions [3], a Boltzmann equation for the electron energy distribution function must be coupled to the vibrational kinetics for a realistic description of the plasma evolution.

In this context, electron-molecule collisions, involving vibrationally excited H_2 species, assume a role of primary importance so that any numerical simulation of the sources requires, as basic input data, the knowledge of the related cross sections for a kinetic treatment of the problem.

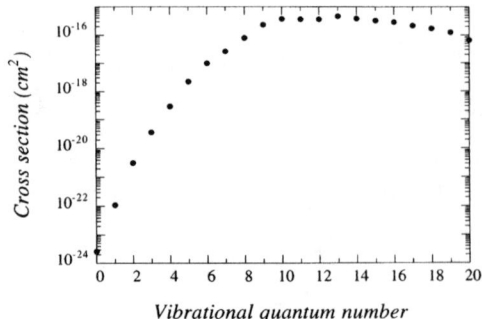

FIGURE 1. Cross section for the process $D_2(X^1\Sigma_g^+, v_i) + e \to D_2^- \to D + D^-$ as a function of initial vibrational level. Each value, indicated by circles, corresponds to the maximum of the "cross section-incident energy" curve.

A lot of electron-impact cross sections have been obtained in the past several years, and complete sets of data are now available for numerous processes. Recently, an attempt to collect and asses all the existing data has been made in our group and a large spectrum of values, as a function of the incident electron energy and vibrational states, are now available for the scientific community [4]. In this paper we will review the cross section database discussing also very briefly the main aspects of the role that electron-molecule collisions play in negative ion plasma chemistry.

2. RESONANT CROSS SECTIONS

Dissociative attachment cross sections for process (1), have been obtained in early calculations for ground state H_2 and D_2 molecules by using the resonant method [5], and recently extended to all the six molecular hydrogen isotopes [4]. New experimental measurements, however, performed in molecular hydrogen gas irradiated by high power excimer laser [6], have shown that dissociative attachment may efficiently occur also from molecules in excited electronic Rydberg states. Evaluations of the rate constants for these processes in fact, lead to estimated values of 3 or more orders of magnitude larger than the corresponding quantities for process (1) [7]. Although the mechanism for this extraordinarily efficient production of negative ions has still not been clarified [8], these experiments have focused the attention on the excitation processes to and among excited electronic states and on their role in plasma kinetics. Electron-impact cross sections for some of these processes are discussed in sections 4 and 5.

Two alternative exit channels for the resonant state H_2^-, leading respectively to dissociation or vibrational excitation, are represented by

$$H_2(X^1\Sigma_g^+, v_i) + e \to H_2^- \to H(1s) + H(1s) + e \qquad \text{dissociation} \qquad (2)$$

$$H_2(X^1\Sigma_g^+, v_i) + e \to H_2^- \to H_2(X^1\Sigma_g^+, v_f) + e \qquad \text{vibrational excitation} \qquad (3)$$

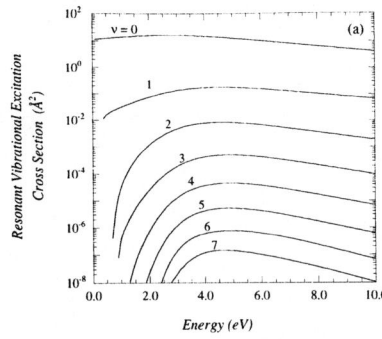 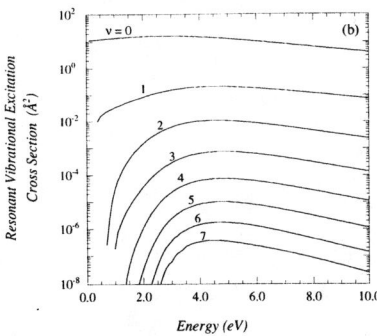

FIGURE 2. Cross section, as a function of energy for different final vibrational levels, for the process (a) $T_2(X^1\Sigma_g^+, v_i=0) + e \to T_2^- \to T_2(X^1\Sigma_g^+, v_f) + e$; (b) $DT(X^1\Sigma_g^+, v_i=0) + e \to DT^- \to DT(X^1\Sigma_g^+, v_f) + e$ [11].

In both cases the molecular ion decays into the neutral molecule, by ejecting the attached electron, reaching either the continuum vibrational spectrum of $b^3\Sigma_u^+$ and $X^1\Sigma_g^+$ states, with subsequent dissociation (process (2)), or the bound vibrational levels of the ground electronic state (process (3)). Cross section for process (2) have firstly been calculated by Atems and Wadehra [9] and more recently discussed by Fabrikant et al. [10], for $v_i = 0, 3, 6, 9, 12$ and for an incident electron energy below 18 eV.

Process (3), which is usually indicated as e-V process, is an indirect vibrational excitation which is mainly induced by low energy electrons and populates the first few vibrational levels. Cross sections have been reported for H_2 molecule by Wadehra [5]. Work is now in progress in our laboratories to extend the calculations to all the six isotopic variants [11]. Preliminary results for T_2 and DT molecules are shown in figure 2.

3. RADIATIVE DECAY PROCESS

The more effective process leading to excitation of high vibrational levels in negative ion plasma conditions, is probably the so-called E-V process, which can be pictured as

$$H_2(X^1\Sigma_g^+, v_i) + e \to H_2^*(singlet\ states) + e \to H_2(X^1\Sigma_g^+, v_f) + e + h\nu \qquad (4)$$

The excitation of the ground-state H_2 molecule to the singlet states (B $^1\Sigma_u^+$ and C $^1\Pi_u$ give the main contribution) is induced by high energy electrons due to the large transition energy, and it is followed by spontaneous emission back to the ground state with the subsequent repopulation of the whole manifold of vibrational levels. Complete sets of cross sections as a function of the incident energy and for all v_i and v_f quantum numbers has been obtained for H_2 molecule by using the impact-parameter method,

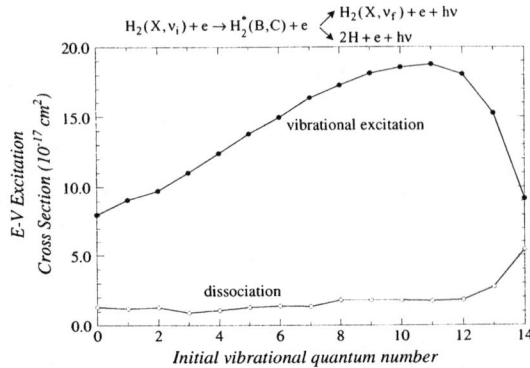

FIGURE 3. Cross section, as a function of initial vibrational quantum number, for the processes: $H_2(X^1\Sigma_g^+, v_i) + e \to H_2^*(\text{excited singlet states}) \to H_2(X^1\Sigma_g^+, v_f) + e + h\nu$ (filled circles); $\to H_2(X^1\Sigma_g^+, \varepsilon) + e + h\nu$ (open circles).

while a limited spectrum of v_i values is available for deuterium [12]. An example for hydrogen is shown in figure 3 (filled circles) for a fixed incident energy of 40 eV.

Dissociation may occur also through process (4) via decay on the repulsive part of the ground state potential curve. In this case two hydrogen atoms are produced according to the scheme

$$H_2(X^1\Sigma_g^+, v_i) + e \to H_2^*(\text{singlet states}) + e \to H_2(X^1\Sigma_g^+, \varepsilon) + e + h\nu$$
$$\hookrightarrow H(1s) + H(1s) \qquad (5)$$

where ε is the vibrational continuum energy. Open circles in figure (3) indicates the calculated cross sections for this channel [12].

4. ELECTRONIC EXCITATION AND DISSOCIATION

Electronic transitions induced by electron impact can bring the molecule to a bound vibrational level of a given excited electronic state (state-to-state excitation) or in the vibrational continuum of the same state with subsequent dissociation. Both these processes depopulate the ground state vibrational levels reducing the possibility of negative ion production via dissociative attachment. De-excitation processes however, such as spontaneous decay from excited electronic states discussed in section 2, can redistribute the vibrational population favouring indirectly the process (1). State-to-state and dissociative cross sections have been calculated, as a function of incident energy and vibrational states by using the impact-parameter method, for the following processes

$$H_2(X^1\Sigma_g^+, v_i) + e \to H_2(B^1\Sigma_u^+, v_f) + e \qquad \text{excitation} \qquad (6)$$

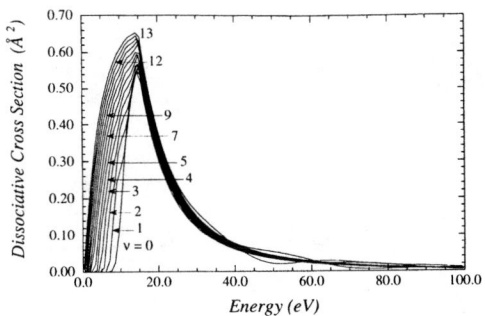

FIGURE 4. Cross sections as a function of energy for the process $H_2(X^1\Sigma_g^+, v_i) + e \to H_2(b^3\Sigma_u^+) \to H + H + e$, for different initial vibrational levels.

$$H_2(X^1\Sigma_g^+, v_i) + e \to H_2(B^1\Sigma_u^+, \varepsilon) + e \to 2H + e \qquad \text{dissociation} \qquad (7)$$

$$H_2(X^1\Sigma_g^+, v_i) + e \to H_2(C^1\Pi_u, v_f) + e \qquad \text{excitation} \qquad (8)$$

$$H_2(X^1\Sigma_g^+, v_i) + e \to H_2(C^1\Pi_u, \varepsilon) + e \to 2H + e \qquad \text{dissociation} \qquad (9)$$

$$H_2(X^1\Sigma_g^+, v_i) + e \to H_2(B', B''^{\,1}\Sigma_u^+; D, D'^{\,1}\Pi_u, v_f) + e \qquad \text{excitation} \qquad (10)$$

$$H_2(X^1\Sigma_g^+, v_i) + e \to H_2(B', B''^{\,1}\Sigma_u^+; D, D'^{\,1}\Pi_u, \varepsilon) + e \to 2H + e \qquad \text{dissociation} \qquad (11)$$

Cross section for processes (6)-(9) are available also for T_2, HD and DT molecules [4, 13], while for all the other cases calculations have been performed only for hydrogen and deuterium [14]. Total cross sections (state-to-state + dissociative) for processes (6) and (8) are available also in form of analytical fits [4]. Vibrational scaling laws have been found for all the excitation processes [4].

The main dissociative channel in hydrogen plasmas is the excitation to the completely repulsive triplet state $b^3\Sigma_u^+$

$$H_2(X^1\Sigma_g^+, v_i) + e \to H_2(b^3\Sigma_u^+, \varepsilon) + e \to 2H + e \qquad (12)$$

Complete sets of cross sections have been calculated for H_2 and D_2 molecules by using the classical Gryzinski method [4]. Figure 4 shows the cross section curves, plotted against the incident energy, for all the initial vibrational levels.

5. EXCITED-TO-EXCITED STATE ELECTRONIC TRANSITIONS

State-to-state and dissociative cross sections for transitions among excited electronic states have been obtained by using the impact-parameter method for the processes

$$H_2(B^1\Sigma_u^+, v_i) + e \to H_2(I^1\Pi_g, v_f) + e \qquad (13)$$

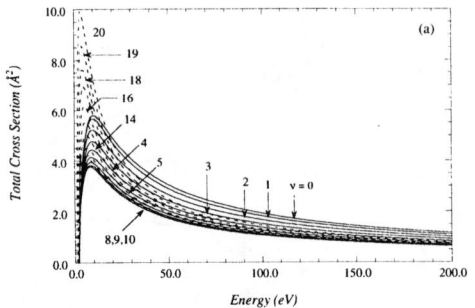
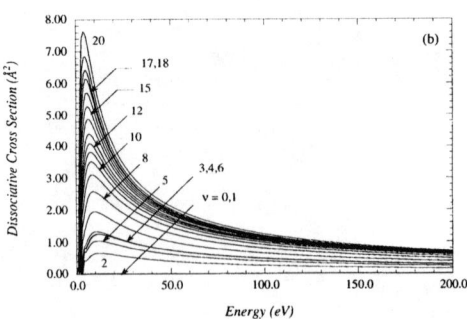

FIGURE 5. Cross sections as a function of energy for the process (a) $H_2(B^1\Sigma_u^+, v_i) + e \rightarrow H_2(I^1\Pi_g) + e$; (b) $H_2(B^1\Sigma_u^+, v_i) + e \rightarrow H_2(I^1\Pi_g) + e \rightarrow H + H + e$.

$$H_2(a^3\Sigma_g^+, v_i) + e \rightarrow H_2(d^3\Pi_u, v_f) + e \tag{14}$$

$$H_2(c^3\Pi_u, v_i) + e \rightarrow H_2(g^3\Sigma_g^+, v_f) + e \tag{15}$$

$$H_2(c^3\Pi_u, v_i) + e \rightarrow H_2(h^3\Sigma_g^+, v_f) + e \tag{16}$$

Complete spectrum of total and dissociative vibrational cross sections for process (13), shown in figure 5, can be found in ref. [4].

Calculations for the triplet-triplet transitions (14)-(16) have been recently performed by using the impact parameter method [15]. Total and dissociative cross sections for the first two processes are shown in figures 6 and 7.

A particular treatment was required in the last case, due to a potential barrier above the dissociation limit in the potential curve for the h state (figure 8b). The presence of a barrier implies the existence of quasi-bound states (resonant states) which can either result in dissociation, by tunneling effect through the barrier or, if they live long enough, fall back to the lower c state.

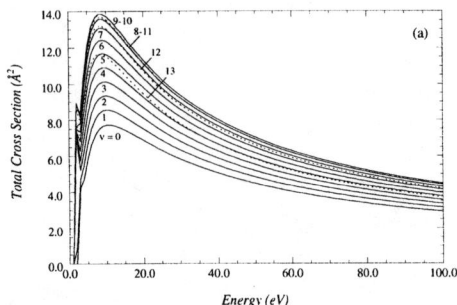
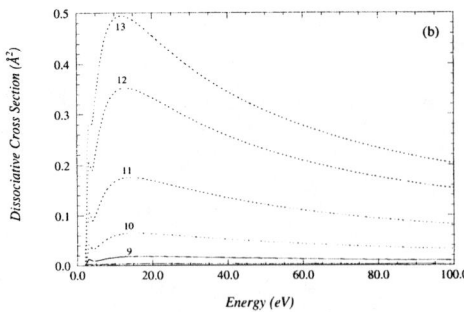

FIGURE 6. Cross sections as a function of energy for the process (a) $H_2(a^3\Sigma_g^+, v_i) + e \rightarrow H_2(d^3\Pi_u) + e$; (b) $H_2(a^3\Sigma_g^+, v_i) + e \rightarrow H_2(d^3\Pi_u) + e \rightarrow H + H + e$.

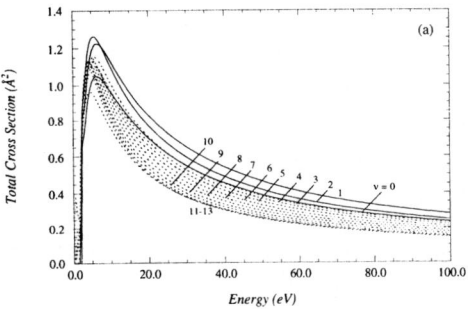
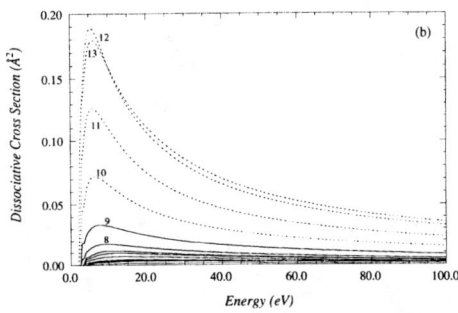

FIGURE 7. Cross sections as a function of energy for the process (a) $H_2(c^3\Pi_u, v_i) + e \to H_2(g^3\Sigma_g^+) + e$; (b) $H_2(c^3\Pi_u, v_i) + e \to H_2(g^3\Sigma_g^+) + e \to H + H + e$.

The quasi-bound states have been searched by evaluating the so-called internal amplitude (IA) defined as

$$IA(\varepsilon) = \int_{R_a}^{R_b} \frac{|\Psi_\varepsilon(R)|^2 \, dR}{[R_b - R_a]} \tag{17}$$

where ε is the continuum vibrational energy, $\Psi_\varepsilon(R)$ the corresponding wavefunction and R_a and R_b the left and right classical turning points respectively for the considered vibrational state.

In figure 8a is shown a plot of the internal amplitude, against the continuum energy, displaying three peaks of resonance (the small picture shows the second peak at a

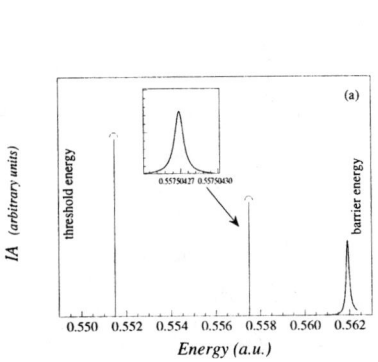
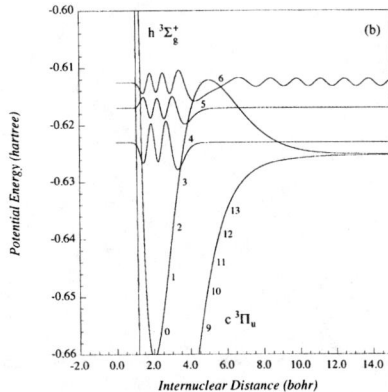

FIGURE 8. (a) Internal amplitude (17) as a function of continuum energy (the second peak is magnified in the inserted picture); (b) H_2 potential energy curves for $c^3\Pi_u$ and $h^3\Sigma_g^+$ electronic states and quasi-bound vibrational wavefunctions for three different energies corresponding to the peaks of figure 8a.

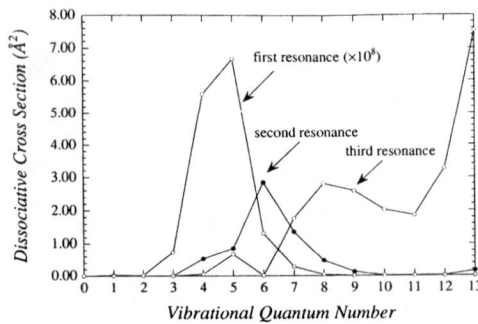

FIGURE 9. Quasi-bound state dissociative cross section as a function of initial vibrational quantum number.

closer resolution) inside which quasi-bound states exist. In figure 8b are shown some of the quasi-bound vibrational wave functions for three different values of the continuum energy belonging, correspondingly, to the three resonances.

The dissociative cross sections for these states, calculated by evaluating the tunneling probability, are shown in figure 9. The cross sections for the excitation to the first region of resonance display very small values, which implies that the corresponding quasi-bound states behave like real bound states; on the contrary, second and third resonances give a non-vanishing contribution to dissociation in particular for intermediate and high vibrational levels respectively.

Figure 10 finally, shows the total and dissociative cross sections for $c \rightarrow h$ transition which include also the contribution coming from the resonant vibrational states.

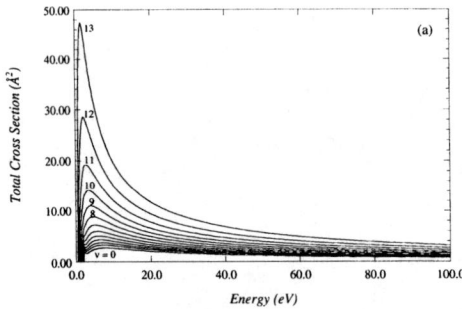
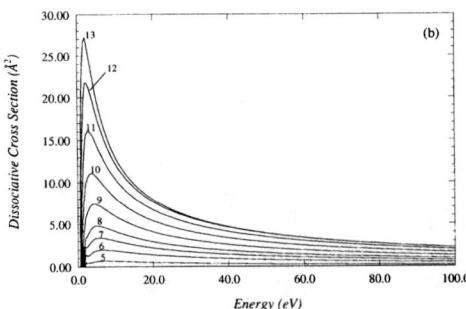

FIGURE 10. Cross sections as a function of energy for the process (a) $H_2(c^3\Pi_u, v_i) + e \rightarrow H_2(h^3\Sigma_g^+) + e$; (b) $H_2(c^3\Pi_u, v_i) + e \rightarrow H_2(h^3\Sigma_g^+) + e \rightarrow H + H + e$.

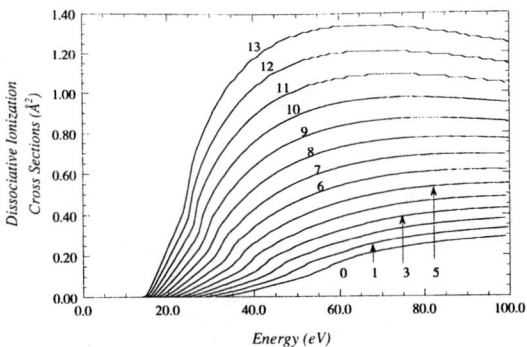

FIGURE 11. Cross section for the process $H_2(X^1\Sigma_g^+, v_i) + e \rightarrow H_2^+(^2\Sigma_u^+, \varepsilon) + 2e \rightarrow 2H + e$ as a function of energy for different initial vibrational levels.

6. IONIZATION

The Gryzinski method has been applied to the calculation of direct ionization cross sections for the processes [4]

$$H_2(X^1\Sigma_g^+, v_i) + e \rightarrow H_2^+(X^2\Sigma_g^+, v_f) + 2e \qquad \text{ionization} \qquad (18)$$

$$H_2(X^1\Sigma_g^+, v_i) + e \rightarrow H_2^+(X^2\Sigma_g^+, \varepsilon) + 2e \rightarrow 2H + e \qquad \text{dissociative ionization} \qquad (19)$$

$$H_2(X^1\Sigma_g^+, v_i) + e \rightarrow H_2^+(^2\Sigma_u^+, \varepsilon) + 2e \rightarrow 2H + e \qquad \text{dissociative ionization} \qquad (20)$$

Process (18) and (19) describe the ionization through the bound part of the ground state potential curve of H_2^+ ion and the continuum branch respectively; process (20) takes place by ionization through the completely repulsive state $^2\Sigma_u^+$ of the molecular ion. Cross sections for this last process are shown in figure 11.

ACKNOWLEDGMENTS

The authors thank Prof. M. Capitelli for useful discussions. The present work has been supported by ASI (I|R|038|01) and MIUR (cof. 2001 Project no.2001031223_009).

REFERENCES

1. M.Capitelli, R.Celiberto, and M.Cacciatore, in *"Advanced in Atomic, Molecular and Optical Physics: Cross Section Data"*, edited by M. Inokuti, Academic Press, N.Y. and London, **33**, 1994, pp. 321-372.
2. J.N.Bardsley and J.M.Wadehra, "Dissociative attachment and vibrational excitation in low-energy collisions of electrons with H_2 and D_2", *Physical Review A* **20**, 1398-1405 (1979).
3. C.Gorse, R.Celiberto, M.Cacciatore, A.Laganà and M.Capitelli, "From dynamics to modeling of plasma complex systems: negative ion (H^-) sources", *Chemical Physics* **161**, 211-227 (1992).

C.Gorse, M.Bacal, R.Celiberto and M.Capitelli, "Nonequilibrium plasma kinetics of D_2 in magnetic multicusp plasmas for (D^-) production", *Chemical Physics Letters* **192**, 161-165 (1992).

4. R.Celiberto, R.K.Janev, A.Laricchiuta, M.Capitelli, J.M.Wadehra and D.E.Atems, "Cross section data for electron-impact inelastic processes of vibrationally excited molecules of hydrogen and its isotopes", *Atomic Data and Nuclear Data Tables* **77**, 161-213 (2002).
5. J.M.Wadehra in *"Nonequilibrium Vibrational Kinetics"*, edited by M. Capitelli, Springer-Verlag, New York and London, 1986.
6. L.A.Pinnaduwage, L.G.Christophorou, "H^- formation in laser-excited molecular hydrogen", *Physical Review Letters* **70**, 754-757 (1993).
7. P.G.Datskos, L.A.Pinnaduwage, "Photophysical and electron attachment properties of ArF-excimer-laser irradiated H_2", *Physical Review A* **55**, 4131-4142 (1997).
8. M.Capitelli, R.Celiberto, A.Eletskii, A.Laricchiuta, "Electron-molecule dissociation cross sections of H_2, N_2 and O_2 in different vibrational levels", *Atomic and Plasma-Material Interaction Data for Fusion* **9**, 47-64 (2001).
9. D.E.Atems and J.M.Wadehra, "Non local effects in dissociative electron attachment to H_2", *Physical Review A* **42**, 5201-5207 (1990).
10. I.I.Fabrikant, J.M.Wadehra and Y.Xu, in "Atomic and Molecular Processes in Divertor Plasma Volume Recombination", edited by R.K.Janev and D.R.Schultz, *Physica Scripta* **T96**, 45-51 (2002).
11. R.Celiberto et al., unpublished results.
12. R.Celiberto, M.Capitelli, U.T.Lamanna, "Vibrational excitation of $H_2(X^1\Sigma_g^+, v)/D_2(X^1\Sigma_g^+, v)$ through excitation of electronically excited singlet states and radiative decay", *Chemical Physics* **183**, 101-106 (1994).
13. R.Celiberto, R.K.Janev, and A.Laricchiuta, "Total and dissociative electron-impact cross sections for $X^1\Sigma_g^+ \to B^1\Sigma_u^+$ and $X^1\Sigma_g^+ \to C^1\Pi_u$ transitions of vibrationally excited tritium and deuterium-tritium molecules", *Physica Scripta* **64**, 26-33 (2001).
14. R.Celiberto, A.Laricchiuta, U.T.Lamanna, R.K.Janev and M.Capitelli, "Electron-impact excitation cross sections of vibrationally excited $X^1\Sigma_g^+$ H_2 and D_2 molecules to Rydberg states", *Physical Review A* **60**, 2091-2103 (1999).
15. R.Celiberto et al., unpublished results.

Influence of Wall Material on VUV Emission from Hydrogen Plasma in H⁻ Source

M. Bacal[a], M. Glass-Maujean[b], A.A. Ivanov Jr[a], M. Nishiura[c], M. Sasao[d], M. Wada[e]

[a] *LPTP, Ecole Polytechnique, UMR 7648 du CNRS, 91128 Palaiseau, France*
[b] *DIAM, Université Paris VI, UPMC/CNRS, 4, Place Jussieu, Tour 12, BP75, 75252 Paris, France*
[c] *RIKEN, Wako, Saitama 351-0198, Japan*
[d] *NIFS, Toki, Gifu 509-5292, Japan*
[e] *Doshisha University, Kyotanabe, Kyoto 610-0321, Japan*

Abstract. The study of VUV emission from a hydrogen plasma produced in a filament discharge in a magnetic multicusp device showed that the use of tantalum and tungsten filaments leads to significant differences in the spectra. The effect of the filament material is interpreted in terms of the fresh film of this material, deposited on the wall. The synthetic spectrum convoluted with our apparatus function for the conditions of this experiment (gas temperature 500 K, electron energy 100 eV) agrees roughly well with the spectrum obtained with tungsten covered walls, but not with the spectrum obtained with tantalum covered walls. We show that in the case of tungsten covered walls the E-V singlet excitation is indeed a two-step Franck-Condon transition, going through either B or C state from an initial H_2 molecule with v''=0, added to a Franck-Condon transition to highly excited states cascading to the B or C states. The excitation process to high v'' states in the case of tantalum covered walls is a three step process, in which the first step is the formation by recombinative desorption on the wall of a vibrationally excited molecule with v''=1 or 2, which serves as the initial molecule in the subsequent E-V excitation through the B state. The results indicate a larger recombination coefficient of atoms on the tantalum covered wall.

INTRODUCTION

The effect of wall material on the extracted H⁻ ion current has attracted the interest since the discovery of H⁻ volume production. Two different experimental methods were used. Leung et al [1] and Fukumasa et al [2, 3] studied the effect of wall material by installing liners made of different metals on the chamber wall, while plasma was produced by primary electrons emitted from tungsten filaments. Aluminum and copper were found to generate the highest H⁻ current while stainless steel produces the lowest one. The tungsten and tantalum liners studied in [1] gave results between those of aluminum and stainless steel, with tungsten giving a slightly higher yield than tantalum. Fukumasa and Saeki [2] indicated that the difference in H⁻ yield between aluminum, copper and stainless steel liners depends strongly on the pressure of hydrogen gas.

Inoue et al [4] modified the wall material by evaporating filament material on it. In this case the wall was covered with fresh metal film, either tungsten or tantalum. The extracted H⁻ current was larger when tantalum filaments were used, compared to tungsten filaments. These observations were confirmed by work with TRIUMF [5] H⁻ ion sources. Inoue et al [4] suggested that the observed difference in H⁻ yield was due to the higher hydrogen atom recombination coefficient, γ, on tantalum covered wall, via the reduction of the density of atomic hydrogen, which was considered as a poison for volume H⁻ production.

The production by recombinative desorption (RD) of vibrationally excited molecules with higher v'' (up to v''= 9) in the case of tantalum, compared to tungsten (up to v''= 7), as reported by Hall et al [6], seemed also an attractive explanation. However the modeling of a volume H⁻ source [7] showed that the production rate of vibrationally excited molecules by RD for v'' > 2 was low compared to the production due to other important mechanisms known to lead to vibrational excitation in hydrogen plasma, namely :

*radiative decay from singlet electronic states excited by collisions of ground-state molecules with energetic primary electrons, E-V, [8]:

$$e + H_2(X\ ^1\Sigma_g^+\ v'' = 0) \rightarrow e + H_2^*(B\ ^1\Sigma_u^+, C\ ^1\Pi_u\ ...) \qquad (1)$$

$$H_2^*(B\ ^1\Sigma_u^+, C\ ^1\Pi_u\ ...) \rightarrow H_2(X\ ^1\Sigma_g^+\ v'') + h\nu \qquad (2)$$

and

*collisions of ground state molecules with low energy electrons, e-V, through the H_2^- resonance, [9]:

$$e + H_2(v''= 0) \rightarrow H_2^- \rightarrow e + H_2(v'' + \Delta v'') \qquad (3)$$

Therefore RD could not explain the enhanced H⁻ production by dissociative electron attachment, since higher v'' (v'' > 5) are required for efficient H⁻ production by this mechanism. Modeling in [7] showed that RD can be the dominant process only in the net production of molecules with v''≤ 2 in high density plasmas.

Our present work shows for the first time that the enhancement due to RD of the density of molecules with low v'' (possibly v''≤ 2) is important since these molecules can serve as the initial species in the E-V excitation through the singlet state B to higher v''. The cross section for excitation to the singlet states B goes up with the level v'' of the initial molecule [10, 11], therefore the production of high v'' molecules by E-V singlet excitation is enhanced in this scenario. This conclusion was obtained by studying the VUV emission spectrum from the hydrogen plasma in an H⁻ ion source in the range 90 to 170 nm.

Previously Graham [12] studied the VUV spectrum from an H_2 discharge produced by an electron beam in a Pyrex device to estimate the production rate of molecules with v'' ≥ 5. He thus demonstrated the possibility of volume production of negative ions in a device, similar to that in which volume production was first observed. However a detailed analysis of this spectrum has not been carried out.

EXPERIMENTAL SETUP

The schematic diagram of our experimental setup is shown in Fig. 1. The multicusp ion source consists of a cylindrical stainless-steel vessel 27 cm long and 21 cm in diameter. Since Inoue et al [4] have shown that the filament material affected the negative ion density via the film of filament material deposited on the wall, we will denote the situation when tungsten filaments are used, as one with 'tungsten covered wall', and the situation when tantalum filaments are used as one with 'tantalum covered wall'.

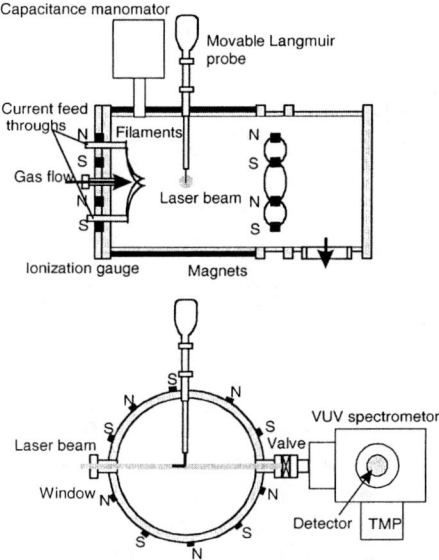

FIGURE 1. Schematic presentation of the experimental setup.

Ten rows of ferrite magnets are applied on the cylindrical wall to produce the multicusp magnetic field. Four magnets close the magnetic field on each end of the vessel. The vessel wall serves as the anode of the discharge. Two filaments (tungsten or tantalum) 100 mm long and 0.5 mm in diameter are located in the multicusp magnetic field next to the end plate. This field acts as a magnetic filter. The gas pressure is monitored by a Baratron gauge. A tungsten cylindrical probe, 10 mm long and 0.1 mm in diameter, was used for measuring the plasma parameters. In particular it was used in the photodetachment measurement of the negative ion density using a Nd-YAG laser. The discharge voltage was 100 V, the discharge current 3 A, the hydrogen pressure 0.1 Pa.

The new filaments are annealed increasing the filament voltage with steps of 0.5V/minute, and up to 80% of the operation voltage. After finishing the annealing of the filaments, we keep the filament heating about one hour. Then we start the discharge about 30 minute before starting the experiment.

The VUV-visible spectrophotometer ($30 \leq \lambda \leq 500$ nm, 0.2 m focal length, Acton Research VM502, with a nominal resolution of 0.4 nm) has been connected to the plasma chamber through the flange where the laser exit window was usually located in the photodetachment experiments. In order to detect the VUV radiation we used a film of sodium salicylate deposited on glass to convert this radiation into visible one. To detect the converted light, we used a photomultiplier HAMAMATSU 721-01. The VUV spectrum was obtained in photon counting mode of the photomultiplier.

The recorded lines of atomic hydrogen could be fitted by a Gaussian. The FWHM for the Ly α line was 0.18 nm. We calibrated the wavelength using the Ly α (121.567 nm) and Ly γ (97.253 nm) lines of atomic hydrogen. The response of the spectrometer went down for wavelength below 120 nm.

We compare the experimental spectra with the synthetic spectrum convoluted by Dr. Xianming Liu [14] with our apparatus function, for our plasma conditions (gas temperature 500 K [15] and electron energy 100 eV). The synthetic spectrum is generated assuming that 100 eV electrons collide with hydrogen molecules in the state v''= 0. This spectrum does not contain the lines of atomic hydrogen, but only molecular lines, from direct B-X and C-X transitions and cascade transitions.

RESULTS AND DISCUSSION

The full VUV emission spectrum recorded with tungsten and tantalum covered walls is shown in Fig. 2 of [13]. Each spectrum is normalized to the corresponding Ly α line amplitude, which is 16% higher when the tantalum wall is used. The aging of sodium salicylate cannot explain this difference, since the spectrum measurement with tungsten wall was carried out five days before that with tantalum wall. Here we compare in Fig. 2 and 3 the experimental spectra obtained with tantalum covered wall (in red) and tungsten covered wall (in blue) with the synthetic spectrum (in green) in two spectral regions: 109nm to 126nm (denoted range A and shown in Fig. 2) and 153 nm to 170 nm (denoted range B and shown in Fig. 3). The reason for the choice of these two ranges, is that the range A is dominated by C-X transitions (Werner system) while the range B is dominated by B-X transitions. The resolution of our instrument allows to resolve only imperfectly the rotational lines of a band. In the spectra one can see the rotational envelope of a band. Before analyzing the detailed observations in each of these two ranges, let us make two general remarks.
1. The synthetic spectrum roughly reproduces the spectrum measured with tungsten walls through the whole studied range (90 to 170 nm) with the exception of the range

below 118 nm, where the spectrometer response gradually goes down. The spectrum measured with tantalum walls is different and is not reproduced by the synthetic spectrum. The agreement with the synthetic spectrum found in the case of tungsten covered walls leads us to conclude that in the spectrum obtained with tungsten covered walls the Franck-Condon transitions start mainly from ground state molecules with v'' = 0, as it is assumed in the calculation of the synthetic spectrum. The transitions observed with tantalum walls may be initiated from molecules with higher v''.

2. The hydrogen molecular lines obtained with tantalum walls are sensibly more intense than those predicted by the synthetic spectrum when they are generated by B-X transitions.

Range A. A general remark is that most molecular hydrogen C-X lines emitted in this range by a plasma produced in a vessel with tungsten covered walls (Fig. 2) are more intense, than those emitted from a plasma produced in a vessel with tantalum covered walls. However, there are some exceptions. The main comments on range A (Fig. 2) are as follows:

a) The intensity of hydrogen molecular lines with tungsten covered walls is higher compared to their intensity with tantalum walls for the transitions from the levels v' = 2 to 5 of the C state, such as (2-5), (3-6) (4-7), (3-7), (4-8), (5-9).

This may be related to the fact that the Franck-Condon transition probabilities from X(v'') state promote the C(v' = 1, 2, 3) states with respect to C (v' = 0 and 4) [16]. This would confirm the assumption that in the case of tungsten covered walls the Franck-Condon transitions start mainly from v''=0, while in the case of tantalum covered walls contributions from v''>0 are important and hence the v''=0 population may be reduced accordingly.

b) The intensity is slightly (30%) higher with tantalum covered walls for the C-X transitions (1-3), (0-3) and (1-4).

This could confirm the assumption that with tantalum covered walls, the Franck-Condon transitions with the initial molecule in v''=1 are more numerous. For such molecules the levels C (v' = 0 and v' = 1) are promoted compared to C (v' = 2, 3 and 4) [16].

c) When a line is more intense for tantalum covered walls than for tungsten covered walls in a wavelength range dominated by C-X transitions (as it is the case in range A) this can be due to B-X transitions. This is the case near 110.5 nm with the (4-1) B-X rotational band or near 115.5 nm with the (7-3) and (10-4) B-X rotational bands.

Range B. This range (Fig. 3) is dominated by B-X transitions (Lyman system). The lines generated with tantalum covered walls are two to three times more intense than the lines generated with tungsten covered wall.

a) The B(v' = 3, 4 and 5) states are promoted by v'' = 1 [16]. Thus the presence of intense structures (3-10), (3-11), (4-11), (4-12), (5-10), (5-12) indicates that vibrationally excited molecules v''= 1 possibly serve as initial species in the Franck-Condon transitions to the B state.

b) The B-X (7-13) band originates from v''= 0 and is more intense with tantalum covered wall. It is not clear why this happens.

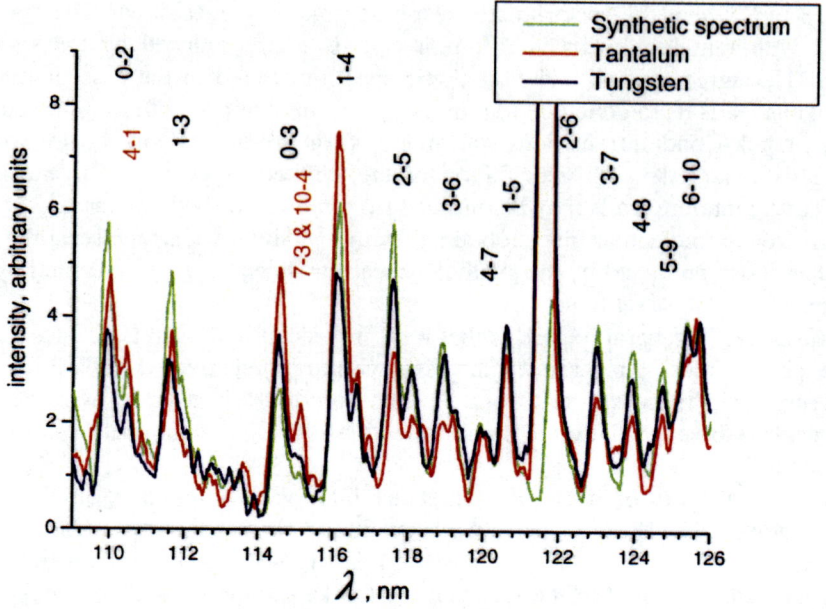

FIGURE 2. Range A spectra. The B-X vibrational transitions are indicated in red, the C-X vibrational transitions are indicated in black.

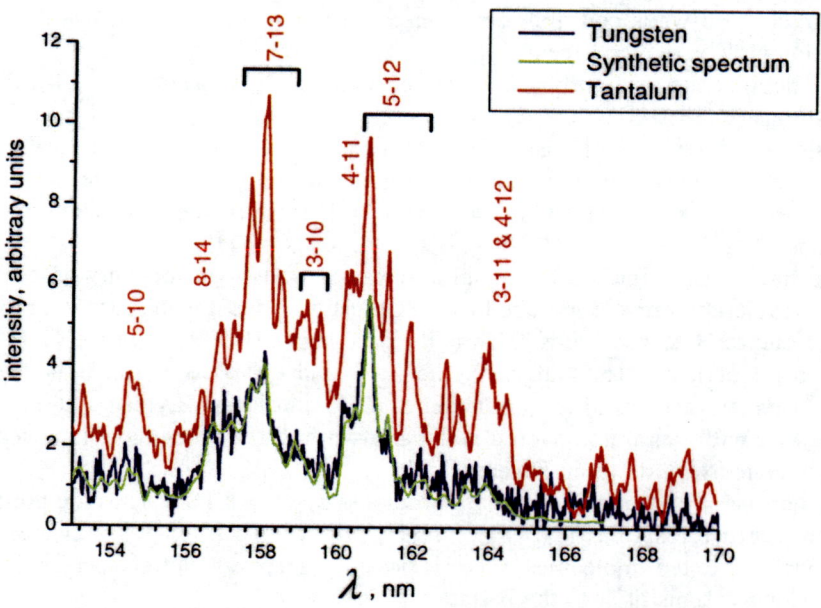

FIGURE 3. Range B spectra. The B-X transitions are indicated in red.

c) The line at 158.5 nm is very intense with tantalum covered walls. It may be generated by the (2-9) B-X band promoted by the v''=1 or 2 initial molecules.

d) Some lines can be amplified in presence of tantalum covered walls also due to cascades, which have as initial state v''>0.

CONCLUSION

1. The VUV spectrum measured from a hydrogen plasma produced in a vessel with tungsten covered walls is roughly reproduced by the synthetic spectrum. Since the latter is based on the assumption that the initial species are in X (v'' = 0) state, we conclude that in the case of tungsten covered walls the E-V singlet excitation is indeed due mainly to Franck-Condon transitions from the v''=0 levels, going through either B or C state or cascading to either B or C state [14]. The gas temperature, 500 K, based on measurements in [15], seems to be correct.

2. The spectrum measured from a hydrogen plasma with tantalum covered walls is different from the synthetic spectrum. The structures in this spectrum which are amplified with respect to the synthetic spectrum are due in most of the cases to B-X transitions, with an initial species in states v''= 1 or 2 (not only v''= 0). The excitation process to high v'' states in this case is not a simple Franck-Condon process as described by Eqs. 1 and 2, but a three step process, in which the first step is the emission by the wall of a vibrationally excited molecule, formed by RD :

$$H + H + \text{wall} \rightarrow H_2 (X, v''= 1 \text{ or } 2) + \text{wall} \quad (4)$$

followed by the E-V excitation of this molecule by Franck-Condon transitions:

$$e + H_2 (X\ ^1\Sigma_g^+\ v'' = 1 \text{ or } 2) \rightarrow e + H_2^* (B\ ^1\Sigma_u^+, C\ ^1\Pi_u \ ...) \quad (5)$$
$$H_2^* (B\ ^1\Sigma_u^+, C\ ^1\Pi_u \ ...) \rightarrow H_2 (X\ ^1\Sigma_g^+\ v'') + h\nu \quad (6)$$

3. Hydrogen atoms have a positive effect in H⁻ production, via the generation of vibrationally excited molecules by RD from the wall, along with their negative effects (H⁻ destruction by associative detachment, quenching of vibrationally excited molecules by V-t transfer).

4. The results indicate a larger recombination coefficient of atoms on the tantalum covered wall. Since the amplification effect of tantalum is related to the generation of low v'' molecules, the large effect observed with tantalum

covered walls implies that the density of these low v'' molecules is much higher for tantalum covered walls, than for tungsten covered walls.

The described phenomenon may be also of interest in the interpretation of the H_2 emission spectra in astrophysics and in the Tokamak divertor physics.

ACKNOWLEDGEMENTS

The support of the European Community (Contract HPRI-CT-2001-50021) is gratefully acknowledged. This work has in part been carried out under the Collaborative Research Program at the National Institute for Fusion Science (Toki, Gifu, Japan). The authors are grateful to R.I. Hall and Xianming Liu for enlightening discussions. We are also grateful to Xianming Liu for calculating the synthetic spectrum for the conditions of our experiment. The collaborations of Françoise Launay, M. Hamabe and H. Yamaoka are gratefully acknowledged.

REFERENCES

1. Leung K.N., Ehlers K.W., Pyle R.V., *Appl. Phys. Lett.*, **47**, 227-228 (1985)
2. Fukumasa O., Saeki S., *J. Phys. D: Appl. Phys.*, **20**, 237-240 (1987)
3. Fukumasa O., Saeki S., Shiratake, *Nucl. Instrum. Methods in Phys. Res.*, **37/38**, 176-179 (1989)
4. Inoue, T., Matsuda, Y., Ohara, Y., Okumura, Y., Bacal, M., Berlemont, P., *Plasma Sources Sci. Technol.*, **1**, 75-81 (1992)
5. Kuo, T., Yuan D., Jayamanna K., McDonald M., Baartman R., Schmor P., Dutto G., *Rev. Sci. Instrum.*, **67**, 1314-1316 (1996)
6. Hall, R.I., Cadez, I., Landau, M., Pichou, F., Schermann, C., *Phys. Rev. Lett.*, **60**, 3-6 (1988)
7. Bacal, M., Capitelli, M., Gorse, C., Skinner, D.A.(1988) « Analysis of Physical Limitations of Volume Negative Hydrogen Ion Production in Tandem Sources » in *Production and Application of Light Negative Ions*, edited by Henk Hopman and Wim van Amersfoort, FOM-Institute for Atomic and Molecular Physics, Kruislaan 407, 1098 SJ Amsterdam, The Netherlands, 1988, pp. 112-117
8. Hiskes, J.R., J. Appl. Phys., **51**, 4592-4594 (1980)
9. Schultz, G.J., Rev. Modern Phys., **45**, 423 (1973)
10. Hiskes, J.R., *J. Appl. Phys.*, **70**, 3409-3417 (1991)
11. Celiberto, R., Capitelli, M., Lamanna, U.T., Chem. Phys., **183**, 101-106 (1994)
12. Graham, W.G., *J. Phys. : Appl. Phys.*, **17**, 2225-2231 (1984)
13. Nishiura, M., Sasao, M., Hamabe, M., Yamaoka, Bacal, M., *Rev. Sci. Instrum.*, **73**, 949-951 (2002)
14. Liu, X., Ahmed, S.M., Multari, R.A., James, G.K., Ajello, J.M., *The Astrophysical Journal Supplement Series*, **101**, 375-399 (1995)
15. Péalat, M., Taran, J.P.E., Bacal, M., Hillion, F., *J. Chem. Phys.*, **82**, 4943-4953 (1985)
16. Allison, A.C., Dalgarno A., *Atomic Data* **1**, 289-304 (1970)

Measurements of H⁻ Density and Work Function of a Plasma Electrode in a Negative Ion Source

M. Nishiura*, M. Sasao[†], and M. Wada[‡]

*The Institute of Physical and Chemical Research, 2-1 Hirosawa, Wako, Saitama 351-0198 JAPAN
[†]National Institute for Fusion Science Toki, Gifu 509-5292, JAPAN
[‡]Doshisha University, Kyotanabe, Kyoto 610-0321, JAPAN

Abstract. The correlation between the work function of the plasma electrode of a H⁻ ion source and H⁻ density was investigated in H_2 plasma, which contains the vapor of an alkali metal (cesium or rubidium). For the work function measurement, the photoelectric current was measured with laser irradiation to the plasma electrode, while the negative ion signal was measured by the laser photodetachment technique simultaneously with an electrostatic probe. As the work function of the plasma electrode was reduced due to alkali metal adsoption, the ratio of H⁻ density to electron density, n_-/n_e, increased gradually. The n_-/n_e ratio became the maximum value when the work function reached the minimum value. The measured work function was well correlated with the n_-/n_e ratio. The dependence of plasma parameters on the discharge current is compared between pure H_2 and H_2 + (Rb/Cs) mixture discharge.

INTRODUCTION

A high current negative ion source is required for a negative ion based neutral beam injector (N-NBI) in thermonuclear fusion, and for an accelerator in high-energy physics. A negative ion source of volume production type can generate an intense beam efficiently and is useful for these applications. This type of ion source is generally divided into two regions by a magnetic filter field. The first is called the driver region, and the second is the extraction region. The former is the production region of vibrationally excited molecules. The latter is that for H⁻ ions to be formed by a dissociative electron attachment reaction. In this region vibrationally excited molecules, $H_2(v'')$, produced by collision with fast electrons in the driver region, encounter slow electrons in a low temperature plasma [1, 2];

$$e + H_2(X^1\Sigma_g^+, v) \rightarrow H^- + H.$$

In order to extract higher beam current at the optimum configuration, the addition of cesium has been a useful method [3], for example, in JT-60U [4], and LHD [5]. It has been reported that the beam intensity of H⁻ tends to be improved, especially at higher discharge power, due to the cesium seeding.

The possible reason for the H⁻ enhancement would be the surface production of H⁻ on the plasma electrode, where the cesium adsorption on the electrode causes a reduction of the work function on the surface, and then the following processes may occur [6],

$H^+ + 2e_{surface} \to H^-$,

$H^0 + e_{surface} \to H^-$.

Shinto et al. [7] have investigated the relation between the work function of the cesium adsorbed molybdenum surface and the extracted H^- beam current using the photoelectron emission method assisted with a laser light, independently.

In this paper we report the results obtained simultaneously by use of the laser photodetachment method, the photoelectron emission method, and the spectroscopic measurement in front of a plasma electrode.

EXPERIMENTAL SETUP

Figure 1 shows an illustration of the experimental setup for the photoelectron current, the photodetachment, and spectroscopic measurements. The 12 rows of Sm-Co magnets for plasma confinement are attached to the surface of the cylindrical vessel (8.5 cm in diameter and 10 cm in length) made of copper. The end flange also has 4 rows of magnets for plasma confinement, two current feedthroughs, and a quartz glass window through which to pass the intense laser light used for photodetachment without causing a reflection and stray light. Hydrogen plasma is generated by two cathode filaments (0.35 mm in diameter, 100 mm in length). The sidewall is biased positively with respect to the cathode filaments to serve as the anode under dc mode operation.

In the extraction region the stainless steel vessel has three ports. Each port is used for a spectroscopic measurement, an electrical probe, and an alkali metal reservoir, respectively. The range of wavelength of the spectroscope (Hamamatsu, PMA-11, which has been improved to measure the spectrum in the near infrared region) is swept from 300 to 900 nm. A couple of Rb lines (1) $5p^2P_{3/2} - 5s^2S_{1/2}$ (2) $5p^2P_{1/2} - 5s^2S_{1/2}$ or Cs lines (1) $6s^2S^{1/2} - 6p2P_{3/2}$ (2) $6s^2S_{1/2} - 6p^2P_{1/2}$ are detected respectively, along with hydrogen Balmer series.

The cylindrical Langmuir probe located in front of the plasma electrode is made of a tungsten wire of 0.35 mm diameter and a tip of 10 mm in length. 1 g of metallic rubidium or cesium is put into the alkali metal reservoir. The temperature of the feedthrough pipe is kept at 270 °C and the reservoir operated around 200 °C, controlled by use of a pair of heaters and thermocouples.

The laser photodetachment technique is used to measure the n_-/n_e ratio. The work function of the molybdenum plasma electrode is measured by using the photoelectron emission method. We use an Ar^+ ion laser (Coherent, photon energy $h\nu = 2.41$ eV, maximum output power $W_L = 3$ W) chopped at a frequency of 2 kHz, f_1, a He-Ne laser (NEC, $h\nu = 1.96$ eV, $W_L = 30$ mW) chopped at a frequency of 1 kHz, f_2. Two lock-in amplifiers detect the photoelectron signals. Two chopped laser beams are mixed and introduced into the ion source. It is irradiated perpendicular to the plasma electrode biased at $V_{pe} = -60$ V. The reflected laser beam on its surface exits from the entrance window in order not to hit other parts of the ion source.

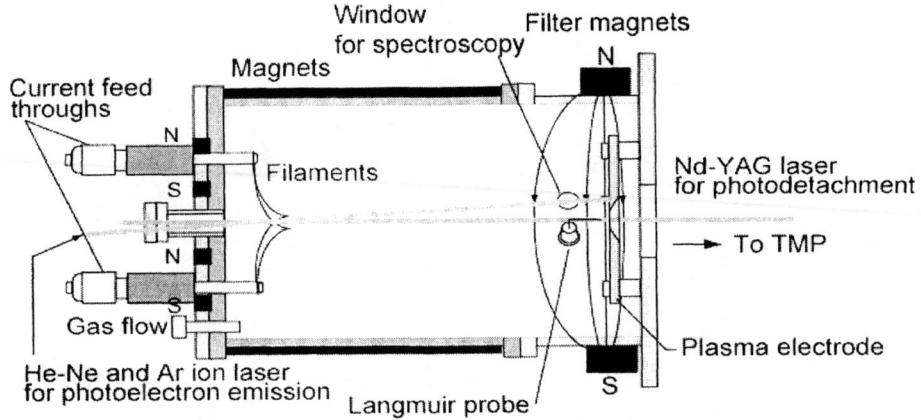

FIGURE 1. Cross sectional view of the configuration for H⁻ density and work function measurements in the multicusp negative ion source.

n_-/n_e RATIO AND WORK FUNCTION

The H⁻ density, the work function of a plasma electrode and the alkali atom spectrum were measured simultaneously by use of the photodetachment method, the photoelectron emission method, and the spectroscopic measurement in front of a plasma electrode. The variation with n_-/n_e ratio and work function is plotted against the intensity of the rubidium spectral line, I_{Rb}, in Fig. 2. The values of the work function are calculated from those of measured photoelectric current [8].

The minimum work function of the Rb adsorbed molybdenum surface is estimated as 1.79 eV, which is near the threshold of the photoelectron emission for the He-Ne laser. We used 1.79 eV for the Rb injection case, because the photoelectric current, I_{pe}, was under the noise level. On the other hand, we determined the work function for the cesium injection case by means of the photoelectron emission using the Ar⁺ ion and the He-Ne laser lights [9]. Before the addition of Rb the n_-/n_e ratio was about 0.4 and the signal of photoelectron current was zero. The n_-/n_e ratio went up slowly with increasing reservoir temperature. The n_-/n_e ratio became the maximum value and the work function reduction reached the peak value. At this time the minimum work function of molybdenum surface adsorbed by rubidium becomes 1.8 eV at $I_{Rb} = 1200$. When I_{Rb} increased further, the work function reaches 2.1 eV. The plasma electrode would be covered with rubidium, because the work function is close to the intrinsic work function of rubidium. We found the n_-/n_e ratio is sensitive to the change of the work function at the surface of the plasma electrode.

From this result, the surface production of H⁻ ions seems to be dominant in H_2 + alkali mixture plasma, since the n_-/n_e ratio reaches the maximum value at the work function minimum. However we can consider that the change of the plasma potential and the n_-/n_e ratio due to the alkali metal addition causes an improvement of the plasma confinement in the driver and extraction region and so influences the extractable H⁻ current.

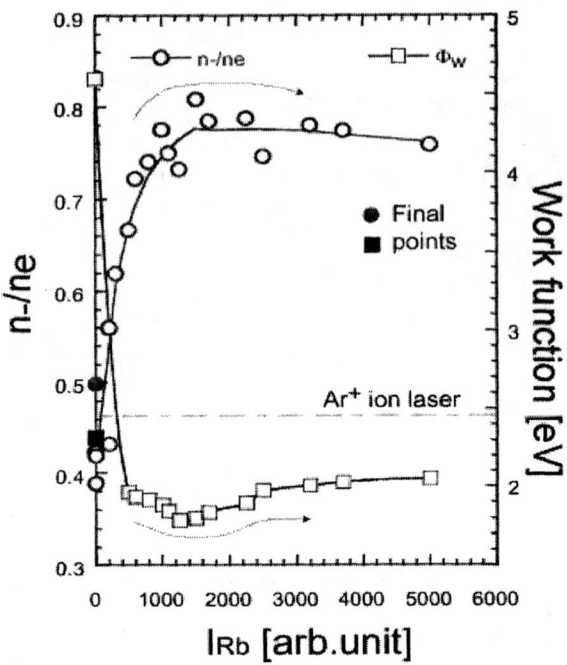

FIGURE 2. Variation of n_-/n_e and work function with the emission intensity of Rb in the extraction region in the H_2 + Rb mixture discharge. $V_{pe} = -60$ V, $W_L = 104$ mW, and $f_1 = 1.14$ kHz.

In both Rb and Cs mixture cases, the maximum value of the n_-/n_e ratio agrees with the point of the minimum work function, although a difference in the H^- enhancement due to the masses of Rb(M = 85.5) and Cs(M = 132.9) does not appear clearly in these experiments. As a large amount of Cs or Rb is introduced into the ion source over the minimum work function, the n_-/n_e ratio decreases gradually. This result is considered to be due to the mutual neutralization between H^- ions and positive ions of alkali atoms, or the collisional electron detachment with alkali atoms.

DEPENDENCE OF PLASMA PARAMETERS ON DISCHARGE CURRENT

The plasma parameters in the extraction region are shown in Figs. 3 and 4. The values of n_e and n_- are of the order of $\sim 10^{11}$ and $\sim 10^{10}$ cm^{-3}, respectively. In pure hydrogen plasma, the difference in the electron density is due to the positioning error of the probe in the magnetic filter field. When constant alkali metal vapor is introduced into the ion source, n_e is reduced to about a half, while n_- increases from 1.1 to 1.7 times higher with the increase of the discharge current, I_d. The addition of alkali metal vapor is effective in the enhancement of H^- at higher discharge current. The plasma potential, V_s, decreases clearly about $1 \sim 2$ eV, while the electron temperature, T_e, does not change or decreases at most about 0.2 eV.

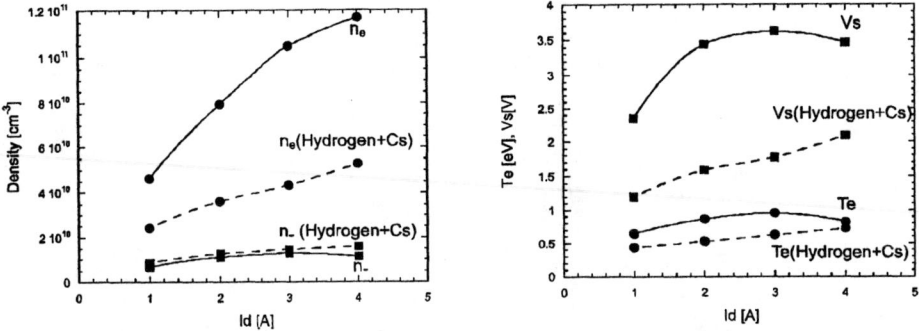

FIGURE 3. Dependence of n_e and n_- on the discharge current for the cases of with (broken) and without Cs (solid) in the left graph. The right graph shows the dependence of T_e and V_s on the discharge current. The hydrogen gas pressure maintained constant P_g of 3.5 mTorr. The discharge voltage was 100 V.

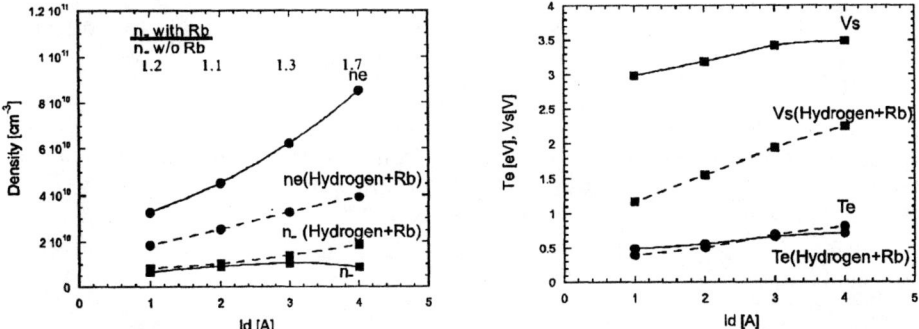

FIGURE 4. Dependence of n_e and n_- on the discharge current for the cases of with (broken) and without Rb (solid) in the left graph. The right graph shows the dependence of T_e and V_s on the discharge current. The discharge was set to the conditions in figure 3. The enhancement ratio of H⁻ density is indicated in the figure.

DISCUSSION

The n_-/n_e ratio is plotted as a function of the electron temperature in Fig. 5. The electron temperature is controlled by the discharge current. At $T_e = 0.45$ eV, the n_-/n_e ratio becomes a maximum whether with or without alkali metal vapor. In pure hydrogen discharge Bacal et al. estimated the n_-/n_e ratio analytically, using the rate equations in plasmas. Employing the same method, we apply it to the results of Fig. 5. When the n_-/n_e ratio corresponds approximately to the cross section for dissociative electron attachment of $H_2(v'')$, this peak at $T_e = 0.45$ eV agrees with the dissociative electron attachment calculated by Wadehara. Since the peak position does not change, the volume production component of H⁻ ions is not influenced by alkali metal vapor, and it seems to increase the absolute value by a factor of $\beta_{sp} \sim 2$ as an effect of alkali metal vapor. Here β_{sp} is defined as the enhancement factor of the n_-/n_e ratio with Cs or Rb. To increase extracted H⁻ current, we should operate and optimize an ion source

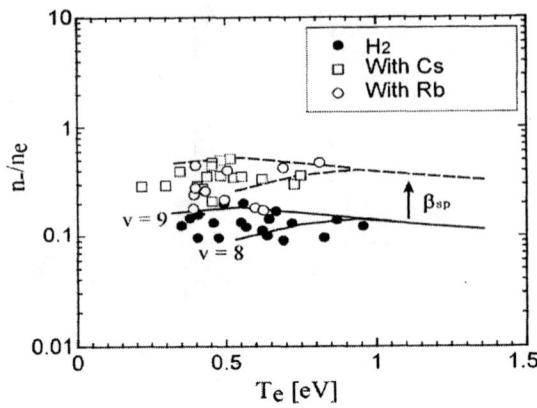

FIGURE 5. Variation of n_-/n_e with the electron temperature in the extraction region in the H_2 discharge. Closed circles are measured data in the pure hydrogen discharge. Open circles and squares represent Rb and Cs seeded cases, respectively. Solid and broken curves indicate the cross sections for dissociative electron attachment of $H_2(v = 8, 9)$ in arbitrary units, calculated by Wadehara [1].

in terms of not only the work function of a plasma electrode, but also the electron temperature cooling due to the addition of alkali metal vapor.

In the future measurements of vibrationally excited molecules of H_2, hydrogen atoms, and molecular ions would be required to understand the production processes of H^- ions in mixture plasmas.

ACKNOWLEDGMENTS

We would like to thank Dr. M. Bacal at Ecole Polytechnique for discussions. One of the authors is also indebted to Dr. T. Katayama at RIKEN for his encouragement throughout this work. Finally, the present work has been carried out under the Collaboration Research Program at N.I.F.S.

REFERENCES

1. J. M. Wadehra and J. N. Bardsley, *Phys. Rev. Letters* **41**, 1795-1798 (1978).
2. M. Bacal, A. M. Bruneteau, W. G. Graham, G. W. Hamilton, and M. Nachman, *J. Appl. Phys.* **52**, 1247-1254 (1981).
3. S. R. Walther, K. N. Leung, and W. B. Kunkel, *J. Appl. Phys.* **64**, 3424-3428 (1988).
4. Y. Ohara, *Rev. Sci. Instrum.* **69**, 908-913 (1998).
5. K. Tsumori, A. Ando, Y. Takeiri, O. Kaneko, Y. Oka, T. Okuyama, H. Kojima, Y. Yamashita, T. Kawamoto, R. Akiyama, and T. Kuroda, *Rev. Sci. Instrum.* **65**, 1195-1197 (1994).
6. K. N. Leung, C. F. A. Van Os, and W. B. Kunkel, *Appl. Phys. Letters* **58**, 1476-1469 (1991).
7. K. Shinto, Y. Okumura, T. Ando, M. Wada, H. Tsuda, T. Inoue, K. Miyamoto and A. Nagase, *Jpn. J. Appl. Phys.* **35**, 1894-1900 (1996).

8. M. Nishiura, M. Sasao, and M. Wada, "Correlation of H^- density with plasma grid work function in a volume production H^- ion source" in *Proceedings of the XII International Conference on Ion Implantation Technology IIT'98*, Kyoto Japan, IEEE transaction, 1999, pp.318-321.
9. Y. Okabe, M. Sasao, H. Yamaoka, M. Wada and J. Fujita, *Jpn. J. of Appl. Phys.* **30**, 1307-1312 (1991).

Study of Isotope Effect on H⁻/D⁻ Volume Production in Low-Pressure H_2/D_2 Plasmas Using VUV Emission

O. Fukumasa, Y. Tauchi, Y. Yabuki, S. Mori and Y. Takeiri*

*Department of Electrical and Electronic Engineering, Faculty of Engineering,
Yamaguchi University, Tokiwadai 2-16-1, Ube 755-8611, Japan
National Institute for Fusion Science, Oroshi, Toki, Gifu 509-5292, Japan

Abstract. Isotope effect on H⁻/D⁻ volume production is studied by using VUV emission. Under the same discharge conditions, VUV emission (corresponding to production of vibrationally excited molecules) in hydrogen plasmas is higher than that in deuterium plasma. Within the present experimental conditions, H⁻ current is higher than D⁻ current, although, considering the factor $\sqrt{2}$, D⁻ ion density in the source is nearly equal to H⁻ ion density under low discharge power. Effect of argon additive in negative ion sources is also discussed briefly.

INTRODUCTION

In a tandem volume source, H⁻ ions are generated by the dissociative attachment of slow plasma electrons e_s ($Te \sim 1$eV) to highly vibrationally excited hydrogen molecules $H_2(v")$ (effective vibrational level $v" \geq 5 \sim 6$). These $H_2(v")$ are mainly produced through singlet excitation H_2^* by fast electrons e_f with energies in excess of 15-20 eV. Namely, H⁻ ions are produced by the following two step process [1, 2]:

$$H_2(X^1\Sigma g, v"=0) + e_f \rightarrow H_2^*(B^1\Sigma u, C^1\Sigma u) + e_f' \quad (1a)$$
$$H_2^*(B^1\Sigma u, C^1\Sigma u) \rightarrow H_2(X^1\Sigma g, v" \neq 0) + h\nu \quad (1b)$$
$$H_2(v") + e_s \rightarrow H^- + H \quad (2)$$

Production process of D⁻ ions is believed to be the same as that of H⁻ ions. To discuss D⁻ production, it is necessary to understand difference in the two step process between H_2 plasmas and D_2 plasmas. For this purpose, we are interested in estimating highly vibrationally excited molecules. The concrete method is to study the production process for $H_2(v")$ / $D_2(v")$ by observing the photon emission $h\nu$ associated with the process (1b), i.e. VUV emission [3, 4].

In this paper, with the use of a double plasma (DP) negative ion source [5], we will study isotope effect of H^-/D^- production in volume negative ion sources. By observing the VUV emission from plasmas, we compare H_2 (v") production with D_2(v") production [6] and discuss further enhancement of H^-/D^- production.

EXPERIMENTAL APPARATUS

Figure 1 shows a schematic diagram of the double plasma type negative ion source [6]. The chamber made of stainless steel is divided by a mesh grid (z = 0 cm) into two regions, i.e. a driver plasma region (the left hand side) and a target plasma region (the right hand side). The target plasma region is a conventional multicusp volume source of negative ions equipped with both a magnetic filter and a plasma grid.

In the driver chamber, argon plasma is produced by arc discharge between hot filaments and the chamber anode, which is grounded. Electrons in argon plasmas are extracted and injected into the target chamber as an electron beam with controlled beam energy eV_B. In the target chamber, hydrogen or deuterium gas is introduced and discharged by the injected electron beam. In this region, the photon emission measurements related to the H_2(v") or D_2(v") production, i.e. the process (1b), are also made by the VUV spectrometer. The spectrometer was normally operated at a resolution of 0.1 nm.

Plasma parameters are measured by Langmuir probes. A magnetic deflection-type ion analyzer is used for relative measurement of the extracted H^- or D^- ions. The plasma grid potential is kept earth potential.

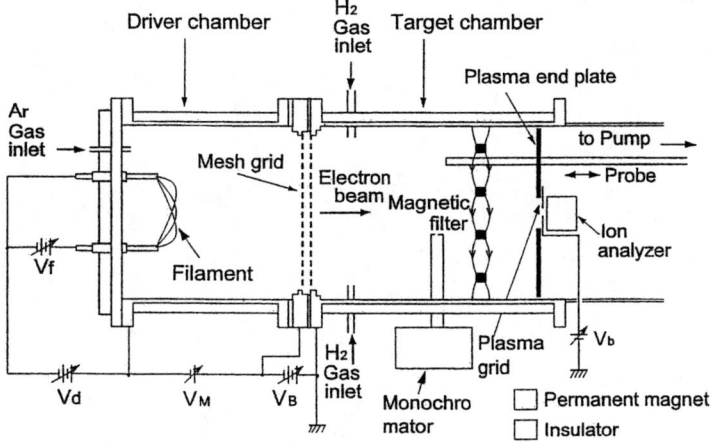

Fig. 1 Schematic diagram of the double plasma type negative ion source.

EXPERIMENTAL RESULTS AND DISCUSSION

Figure 2 shows the typical VUV spectra (a) from the hydrogen and (b) from the deuterium plasmas. Strong intensity (121.8 nm in H_2 plasma and 121.7 nm in D_2 plasma) of VUV spectra is Lyman α of atomic hydrogen and deuterium. According to the numerical simulation [2], $H_2(v" \geq 5)$ are more effective for H^- production. In Fig.2 (a), spectra leading to production of $H_2(v" \geq 5)$ are ranged from 117.5 to 165 nm [3]. Internal energy of molecules $D_2(v" \geq 8)$ are almost the same as that of $H_2(v" \geq 5)$. Therefore, concerning production of vibrational molecules, discussion on VUV spectra from H_2 plasmas can be extend to that from D_2 plasmas. Namely, production of highly vibrationally excited deuterium molecules $D_2(v")$ is related to the emission with the same wavelength range, i.e. 117.5~165 nm. We integrate the VUV spectra (117.5~165 nm) of H_2 and D_2 plasmas to compare $H_2(v")$ production with $D_2(v")$ production. Integration of VUV spectra excludes Lyman-α [6].

Concerning production of H_2 and D_2 plasmas, axial distributions of plasma parameters (i.e. electron density n_e and electron temperature Te) are shown in Fig. 3. Both n_e and Te distributions in D_2 plasmas are slightly higher than ones in H_2 plasmas.

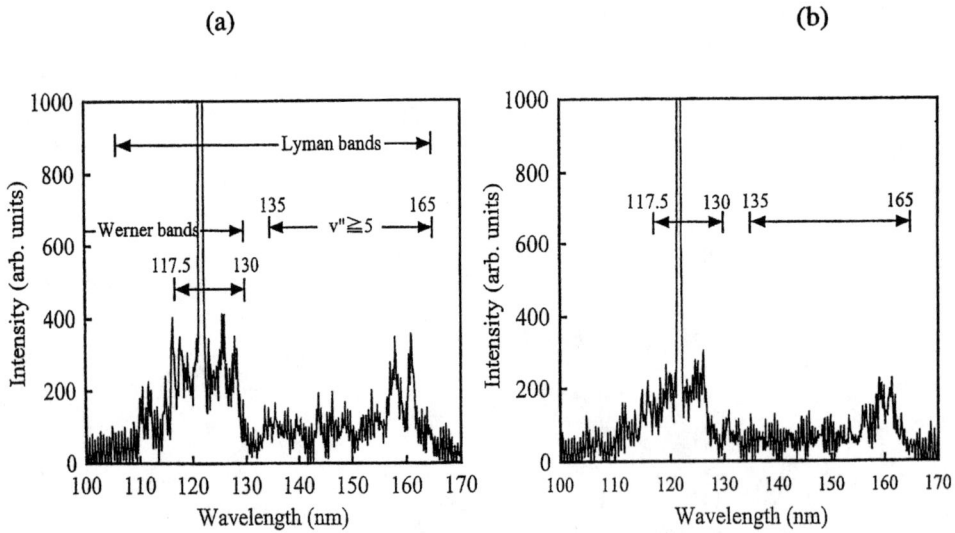

Fig. 2 Typical VUV spectra from (a) hydrogen and (b) deuterium plasmas. Experimental conditions are as follows: discharge voltage of driver plasma $V_d = 40V$, discharge currents $I_d = 4.2A$, $p(H_2$ or D_2 gas$) = 2mTorr$, $p(Ar$ gas$) = 1mTorr$, beam acceleration voltage $V_B = 50$ V and beam current $I_B = 2.0$ A.

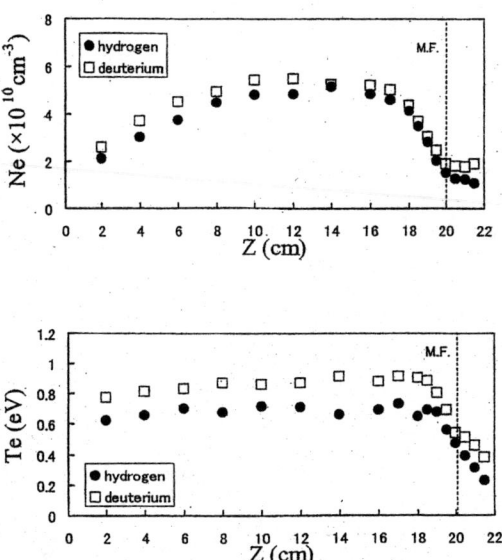

Fig. 3 Axial distribution of plasma parameters (ne and Te). Experimental conditions are as follows: $V_d = 40V$, $V_B = 50V$, $I_B = 3A$, $p(H_2$ or D_2 gas$) = 2$ mTorr and $p(Ar$ gas$) = 1$mTorr.

Figure 4 shows the dependence of negative ion currents on discharge power. D⁻ currents increase almost linearly with discharge power, similar to H⁻ currents. Under the same discharge power, the H⁻ current, I_{H^-}, is higher than the D⁻ current, I_{D^-}. When the value of extraction voltage is the same for both cases, I_{D^-} is usually reduced approximately by factor $1/\sqrt{2}$ compared with I_{H^-} due to mass difference. Although D⁻ density in the source is nearly equal to H⁻ density in low discharge power, the difference between D⁻ and H⁻ densities increases with discharge power.

Figure 5 shows integrated intensities of VUV emissions corresponding to the results in Fig. 4. The ratio of intensity from D_2 plasmas to one from H_2 plasmas keeps nearly the constant value (i.e. 0.7) although intensity from D_2 plasmas is lower than one from H_2 plasmas. On negative ion production, plasma parameter conditions in D_2 plasmas are the same as those in H_2 plasmas. Therefore, reduction of D⁻ density in high power region could be caused by effect of argon additive discussed in Ref. 7. Details are now under study.

Figure 6 shows discharge power dependence of negative ion currents. In this case, plasma is produced in pure H_2 gas or pure D_2 gas. H⁻ current is nearly equal to $\sqrt{2}$ times D⁻ current. Concerning the difference between the results in Fig. 4 and the results in Fig. 6, detailed discussion is necessary and will be continued with measuring H⁻ and D⁻ ion densities in the source.

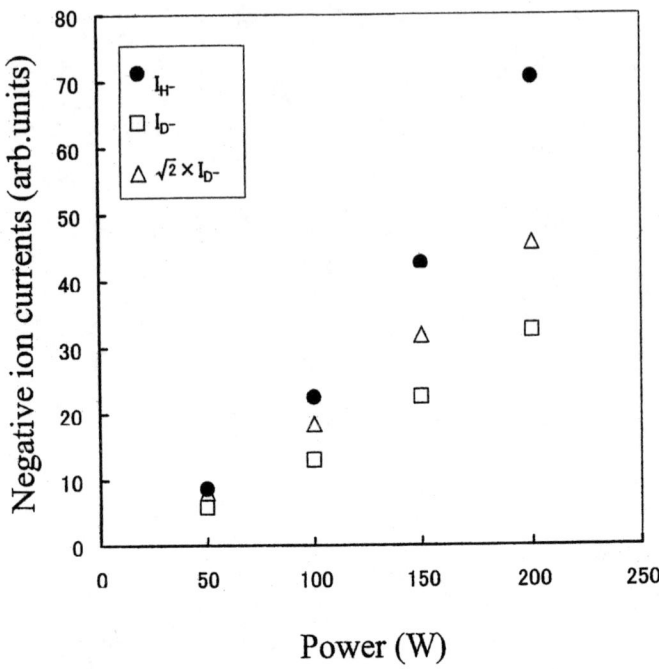

Fig. 4 Dependence of negative ion currents on discharge power. Experimental conditions are as follows: $V_d = 40\text{V}$, $p(\text{H}_2 \text{ or } \text{D}_2 \text{ gas}) = 2$ mTorr, $p(\text{Ar gas}) = 1$ mTorr, $V_B = 50\text{V}$ and extraction voltage $V_{ex} = 600\text{V}$.

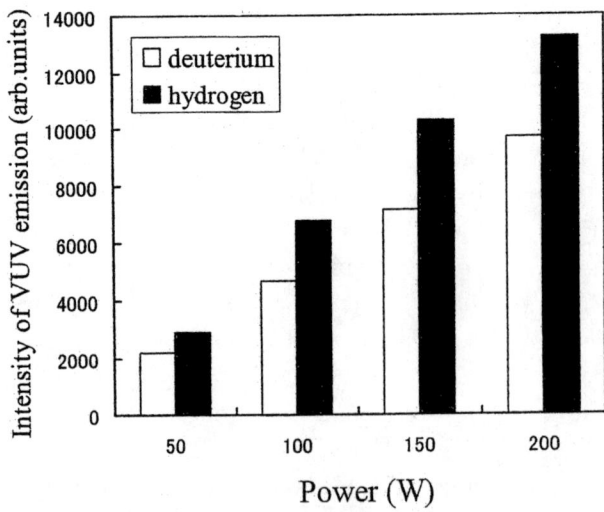

Fig. 5 Integrated intensity of VUV emission spectra from H_2 and D_2 plasmas, corresponding to the results in Fig. 4

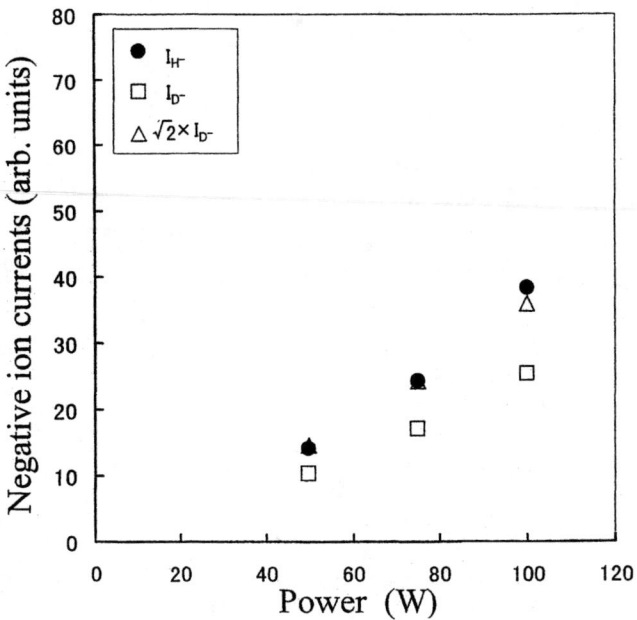

Fig. 6 Dependence of negative ion currents on discharge power. Experimental conditions are the same as ones in Fig. 4 except that argon gas is not injected.

CONCLUSIONS

We have studied isotope effects of H⁻ and D⁻ productions by observing VUV emissions from H_2 and D_2 plasmas. It is found that $D_2(v")$ production is a little less than $H_2(v")$ production. Under the same discharge power, it is confirmed that H⁻ current is higher than D⁻ current. However, considering the factor $1/\sqrt{2}$ due to mass difference, H⁻ and D⁻ productions in the source should be further discussed including measurement of negative ion density in the source and effect of argon additive.

ACKNOWLEDGEMENTS

The authors would like to thank T. Mameda for his assistance of preparing this manuscript. A part of this work was supported by the Grant-in-Aid for Scientific Research from the Ministry of Education, Culture, Sports, Science and Technology, Japan. This work was also carried out as collaboration research of the NIFS.

REFERENCES

1. Hiskes, J. R. and Karo, A. M., *J. Appl. Phys.* **56**, 1927-1938 (1984).
2. Fukumasa, O., *J. Phys.* **D22**, 1668-1679 (1989).
3. Graham, W. G., *J. Phys.* **D17**, 2225-2231 (1984).
4. Fukumasa, O., Mizuki, N. and Niitani, E., *Rev. Sci. Instrum.* **69**, 995-997 (1998).
5. Fukumasa, O. and Iwasaki, T., *Rev. Sci. Instrum.* **65**, 1210-1212 (1994).
6. Yabuki, Y. and Fukumasa, O., in *Proc. Inter. Conf. Phenomena Ionized Gas.* edited by T. Goto, Nagoya University, Nagoya, 2001, Vol. 1, pp. 273-274.
7. Bacal, M., Nishiura, M., Sasao, M., Hamabe, M., Wada, M. and Yamaoka, H., *Rev. Sci. Instrum.* **73**, 903-905 (2002).

SOURCES

A review of recent H⁻ ion source work at Rutherford Appleton Laboratory

J. W. G. Thomason

CLRC RAL, Didcot, Oxon., UK

Abstract. The ion source used at ISIS is a surface plasma source of the Penning type, and routinely produces 35 mA of H⁻ ions during a 200 μs pulse at 50 Hz for uninterrupted periods of up to 50 days. Identical sources have now also been run on both the ISIS RFQ test stand and a dedicated ion source development rig (ISDR). Details of the most recent results from each of these three applications will be presented, along with physical explanations for the observed differences in source behaviour.

INTRODUCTION

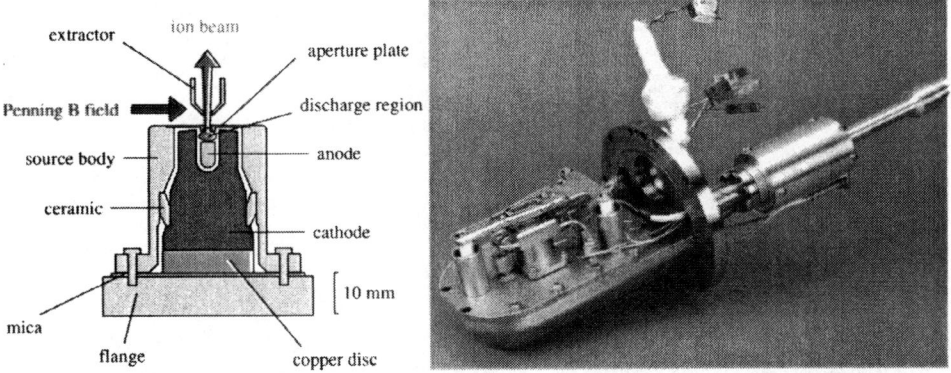

FIGURE 1. a) Schematic of the ISIS H⁻ ion source. b) The ion source assembly.

The design and operation of the H⁻ ion source for the ISIS spallation neutron source at Rutherford Appleton Laboratory (RAL) have previously been described in detail[1,2]. A schematic diagram of the source and a picture of the ion source assembly, which includes electrical feed-throughs, gas supplies and temperature monitors, are shown in figure 1. This source has proved to be extremely successful, and will continue to be used on ISIS when the ageing Cockroft-Walton preinjector accelerator is replaced with a radio frequency quadrupole (RFQ) accelerator. An RFQ test stand has been built at RAL, which is equipped with an identical source to that used on ISIS[3,4]. The standard ISIS ion source will also form the basis of an extensive ion source development programme, carried out on a dedicated rig. This programme is intended to produce improved sources, which will be necessary for next generation projects such as the

European Spallation Source (ESS). The ion source development rig (ISDR) is intended to reproduce, as closely as possible, the beam conditions on either ISIS or the ISIS RFQ test stand, and will also provide an invaluable test bed for new ion source power supplies and other electronics[5].

ISIS RESULTS

Six identical operational ion sources are used in rotation on ISIS, with a complete source change being possible in approximately 2½ hours. The H⁻ ions produced by the ion source traverse a 665 kV DC medium gradient accelerating column to provide a high brightness 35 mA ion beam. Transverse emittance measurements of the 665 keV beam are made using a computer controlled system consisting of in-vacuum beam analysing slits and multi-wire detectors[6], mounted in the low energy drift space (LEDS) between the accelerating column and the ISIS linac.

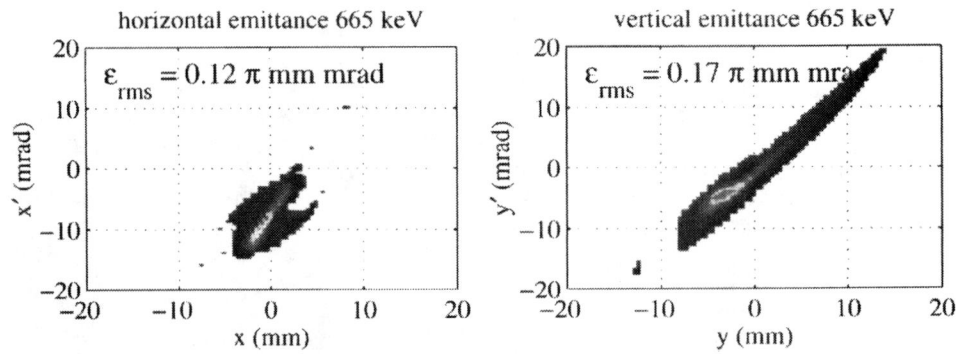

FIGURE 2. Typical ISIS LEDS emittance plots.

Emittance values are taken routinely at the beginning and end of every user cycle on ISIS, and have proved to be remarkably consistent, and independent of which particular ion source is being used. Typical ISIS LEDS emittance plots, with normalised, rms, horizontal and vertical emittance values (ε_H and ε_V) of $\varepsilon_H = 0.12 \pi$ mm mrad and $\varepsilon_V = 0.17 \pi$ mm mrad are shown in figure 2. Emittance measurements closer to the source would give a more sensitive indicator of ion source behaviour, but are not possible on ISIS because of space constraints and the difficulty of operating suitable diagnostic equipment at high potential.

ISIS RFQ TEST STAND RESULTS

Figure 3 shows the ion source, power supplies and other essential services on the RFQ test stand, along with a picture of the RFQ ion source and part of the three solenoid low energy beam transport (LEBT). The ion beam in the LEBT is at 35 keV, and can be measured using two emittance scanners, one in the horizontal and one in the vertical plane, located in a diagnostics box between the second and third solenoids[7]. These are of the slit and collector type and were designed specifically for this application and the ISDR. Each has three axes of travel, with the main drive rapidly moving both the slit and collector into the beam, and then two more drives stepping the slit and collector independently in order to build up an emittance profile of the beam. The associated data acquisition electronics allow the beam to be sampled at 10 µs intervals. Running the beam at 50 Hz enables a full emittance scan to be built up in about 5 minutes.

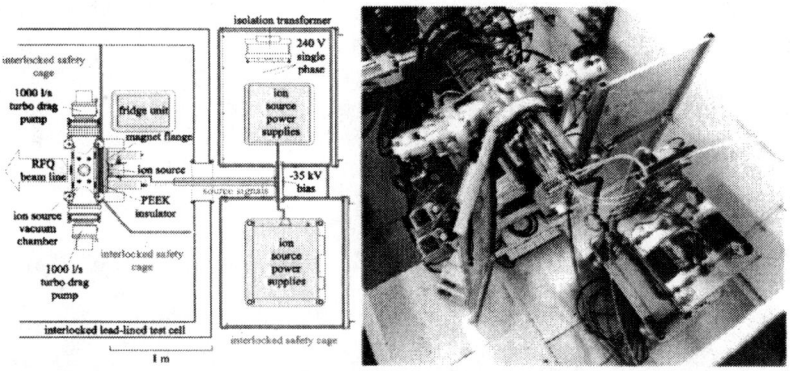

FIGURE 3. a) Schematic of the RFQ ion source layout. b) The RFQ ion source.

The beam current is measured with a toroid immediately after the ion source, and is typically 48 mA, but current measurements downstream indicate that ≈ 36 mA of H⁻ ions are transmitted through the LEBT into the RFQ in a high quality beam pulse. This is comparable with the performance of the ion source on ISIS. The probability is that the current lost in the LEBT is that which, if transmitted, would result in much larger emittance values which would no longer be well matched into the RFQ.

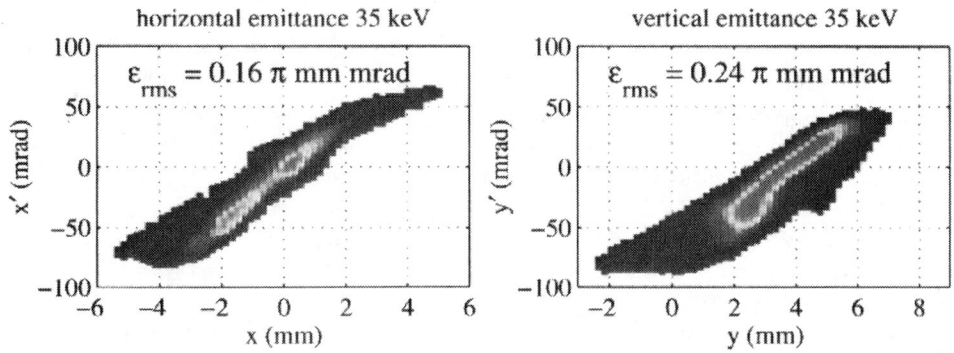

FIGURE 4. Typical ISIS RFQ test stand LEBT emittance plots.

The measured normalised rms emittance values of $\varepsilon_H = 0.16\ \pi$ mm mrad and $\varepsilon_V = 0.24\ \pi$ mm mrad, are shown in figure 4. In addition time resolved emittance measurements have demonstated that, after $\approx 40\ \mu s$ when the beam has been space-charge neutralised, the beam is extremely stable during a given pulse.

ION SOURCE DEVELOPMENT RIG RESULTS

A plan view of the ISDR diagnostics layout is shown in figure 5, alongside a picture of the diagnostics chamber. The ion beam produced on the ISDR is at 35 keV, and can be measured using emittance scanners identical to those on the ISIS RFQ test stand as well as a profile wire[8]. An emittance device using a 'chromox' amorphous scintillator mounted at 45° behind a slit plate is also incorporated. The image cast on the scintillator can be captured with a charge coupled device camera, located outside the vacuum on a glass viewing port. From this image an emittance measurement can be derived from a single beam pulse.

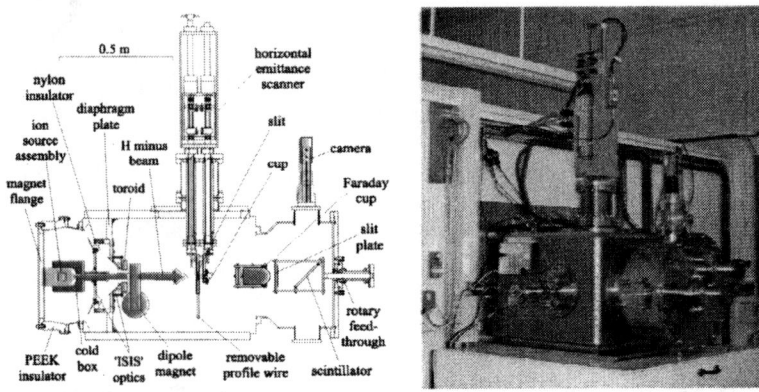

FIGURE 5. a) Schematic plan view of the ISDR diagnostics layout. b) The ISDR diagnostics chamber.

Typically the beam current measured on the ISDR with the toroid immediately after the acceleration optics is 54 mA for a simulation of RFQ beam conditions ('RFQ' optics) and 40 mA for the ISIS conditions ('ISIS' optics). This is simply explained by the fact that the 'ISIS' optics accelerates the beam after it has passed through a 120 mm long × 30 mm diameter tube, which eliminates the possibility of caesium from the ion source reaching the acceleration gap, but also collimates the beam. In the 'RFQ' optics, however, the beam passes through a 30 mm diameter hole in the front of the cold box, and then a 47.5 mm diameter hole in the grounded plate after the acceleration gap, which allows for a greater throughput of current. Beam measurements using the profile wire, 490 mm downstream from the cold box face, show that the horizontal beam width is ≈ 63 mm for the 'ISIS' optics and ≈ 91 mm for the 'RFQ' optics, which demonstrates that some of the additional current may not be commensurate with high brightness.

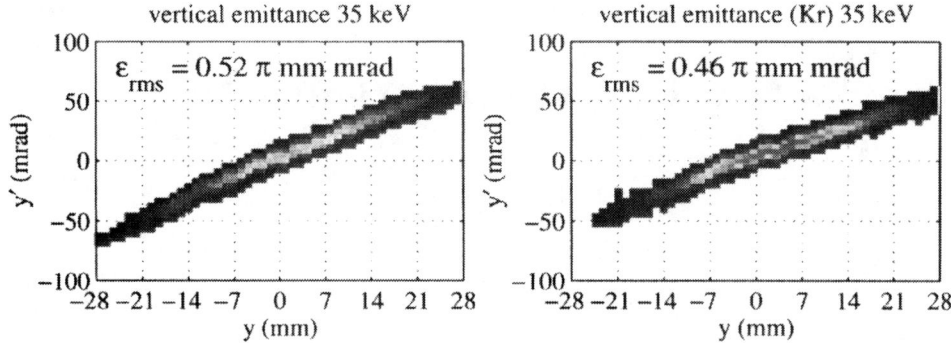

FIGURE 6. ISDR vertical emittance plots without and with krypton.

The residual gas pressure in the ISDR diagnostics chamber is relatively low, leading to sub-optimal space-charge neutralisation in the accelerated H⁻ beam. Preliminary emittance plots, taken using the emittance scanners with the 'ISIS' optics are shown in figure 6, demonstrate the effect of adding a buffer gas to increase space-charge neutralisation. These are both vertical emittances, measured at base pressure (8.0×10^{-6} mbar) and then with krypton gas introduced directly into the diagnostics chamber to raise the pressure to 2.0×10^{-5} mbar. The effects of the buffer gas are immediately obvious. The normalised rms emittance value falls from $\varepsilon_V = 0.52$ π mm mrad to $\varepsilon_V = 0.46$ π mm mrad, the vertical extent of the beam is reduced, the divergence of the beam is lessened and more current is concentrated at the centre of the beam. Whilst these effects are less dramatic than may have been expected, the addition of krypton appears to be entirely beneficial. The emittance values are higher than those measured on the ISIS RFQ test stand, and this probably indicates that the part of the beam being lost in the LEBT on the ISIS RFQ test stand is that which is causing the relatively larger emittances on the ISDR.

ACKNOWLEDGMENTS

Many thanks are due, as ever, to R. Sidlow and M. O. Whitehead for services to ion source physics, to A. P. Letchford and C. P. Bailey for help and advice with all of this work, and M. Perkins and the ISIS injector electrical support section.

REFERENCES

1. R. Sidlow et al., *EPAC 96*, THP084L.
2. J. W. G. Thomason and R. Sidlow, *EPAC 2000*, THP4A07.
3. J. W. G. Thomason et al., *EPAC 2002*, THPRI011.
4. C. P. Bailey et al., *EPAC 2000*, THP4A03.
5. J. W. G. Thomason et al., *Rev. Sci. Instr.* 73(2) p. 896, (2002).
6. M. A. Clarke-Gayther, *EPAC 98, proceedings* p. 1491 (1998).
7. C. P. Bailey et al., *EPAC 2002*, THPLE048.
8. J. W. G. Thomason et al., *EPAC 2002*, THPRI012.

The New HERA H⁻ RF Volume Source and Selected Results of Plasma Investigations

J. Peters

*Deutsches Elektronen-Synchrotron DESY,
Notkestraße 85, 22607 Hamburg, Germany*

Abstract. The HERA H⁻ RF volume source runs now for more than 25000 h without interruption for maintenance. The source delivers 40 mA H⁻ pulses 150μsec long with a repetition of 8 Hz (duty cycle 0.12%). The H⁻ ions are produced without Cs. The status of the source and new plasma investigations will be presented.

INTRODUCTION

At DESY a new source development took place. The internal antenna of the HERA RF H⁻ source was replaced by an external RF coupling. This system has been running for more than 25 000 h without any degradation. This is a significant improvement compared to the 980 h average lifetime of internal antenna systems.

In recent publications (1), (2) the RF coupling problem was raised again. There are six basic reasons why external RF coupling is superior to internal coupling. For details see publication (3).

In order to investigate the starting conditions for beam transport calculations Langmuir measurements were made. These measurements lead to a new model of the plasma transition.

SOURCE DESIGN

Fig. 1 a) shows the RF coil behind the Al_2O_3 ceramic. In Fig.1 b) a view through the filter field magnets into the ceramic cylinder is given.

For the ignition of the source filaments and UV-lamps were tested first. These devices have been seen to have various limitations. Filaments are exposed to the plasma and as a result have a short lifetime. UV-lamps are outside of the plasma and thus can be easily changed but the timing jitter of the H⁻ current is bigger than that with a filament. It was also demonstrated at DESY with internal and external antennas that it is possible to ignite the plasma with an rf frequency which is different from the optimum plasma heating frequency. A frequency source was investigated which switches RF-frequencies. Lasers were not tested because they are bulky and expensive. A spark gap is not feasible in these sources due to the low gas pressure of

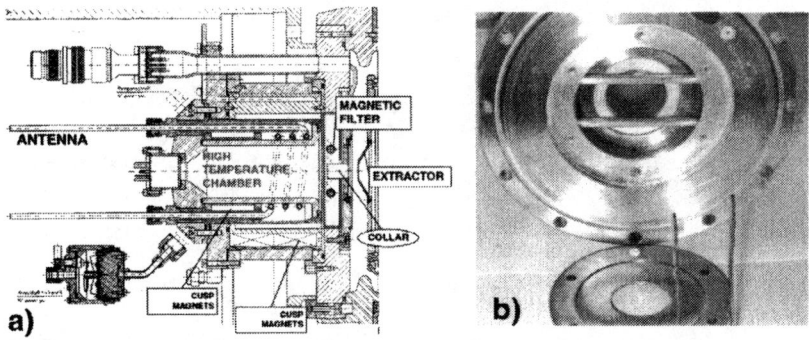

FIGURE 1. a) Drawing of the volume source with external RF coupling and b) a look into the source with the Al_2O_3 ceramic.

about 5 mTorr. According to the Paschen law one would need either a long distance between the electrodes or a very high voltage. This can be circumvented by introducing a tandem source arrangement with an ignition source in the gas pipe connected to the source, using the higher pressure in the pipe for ignition. This source injects then electrons into the main plasma chamber. For details see (4).

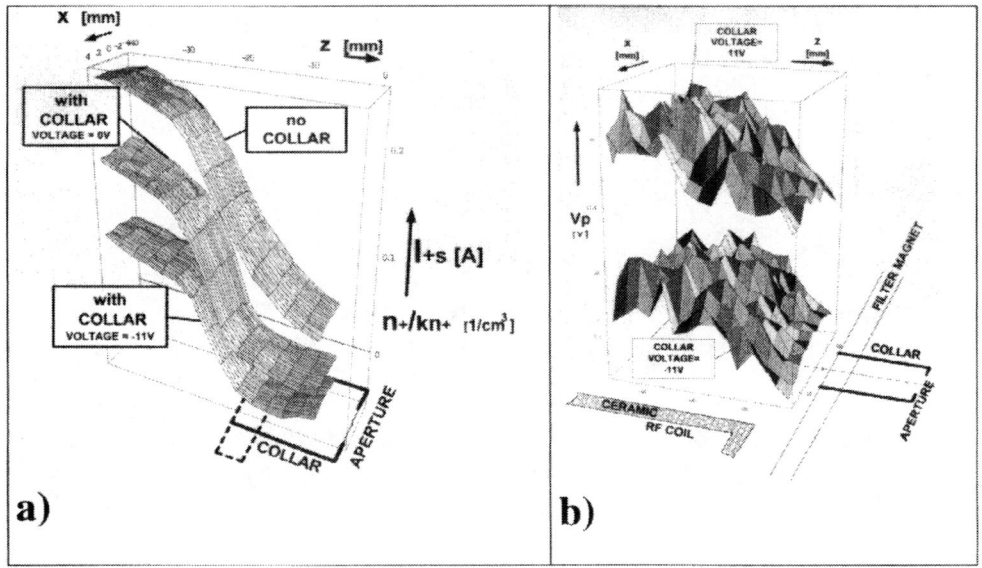

FIGURE 2. a) H⁻ beam without and with a collar bias of 22V measured with a multi faraday cup. b) Plasma potential for a collar biased with +11 and –11 V.

LANGMUIR MEASUREMENTS

A. Langmuir measurement set up and analysis

Langmuir measurements were made with an rf shielded probe 410 mm long and insulated with ceramic. The diameter of the end of the ceramic was 0.8mm and that of the metal tip 0.4mm. It was possible to position the probe in 3D with 3 crossed tables.

B. Plasma density (n), plasma potential (Vp) and plasma temperature

FIG. 2 a) shows the saturation current I_{+s} measured with the Langmuir probe, proportional to the positive particle density (n_+), for a set up without collar and for a collar system with and without bias. It is important to note that I_{+s} becomes negative in a collar system indicating that the plasma deneutralizes and becomes negative. Also plasma potential and plasma temperature were measured and calculated with well known methods. For details see (5). In FIG. 2 b) is shown the measured reduction of Vp with a negative bias voltage compared to a positive one. The curve for zero bias lies between these two curves.

A MODEL OF THE PLASMA TRANSITION WITH COLLAR

The following boundary conditions lead to a model (6), (7) (see Fig. 3) for the plasma transition:

Langmuir measurements (Fig.2a) show that the plasma is neutral or positive in front of the filter field and collar and becomes negative in the collar. Collar measurements (5) indicate that positive ions going to the collar are neutralized mainly at the beginning of the collar. Due to the filter field there are only slow electrons inside of the collar. In the plasma there is charge neutrality. $\rho=0$ leads to $dE/dx=0$ ($\rho = \varepsilon\, dE/dx$) at the edge of the plasma and $E=0$ is necessary for quasi neutrality.

In front of the collar high velocity electrons are lost to the environment and slow electrons pass the filter field into the orifice of the collar and are also lost, leading to $\rho = n_+ - n_- > 0$ and $dE/dx > 0$ in this area. This is also indicated by the plasma potential measurements which show: $E = -dV/dx > 0$ and $\rho = \varepsilon\, dE/dx > 0$.

In the first part of the collar or a little before the charge of the plasma changes from positive to negative. At this point we have again $\rho=0$ and $dE/dx=0$ ($E=const= -dV/dx$, i.e. a turning point for V).

In the deneutralized part of the collar H^- H^0 and H^* are produced out of H^+ and H_3^+ leading to $\rho<0$ and $dE/dx<0$. E falls here to zero, $dV/dx = 0$ resulting in a minimum for V which rises from here to the point of extraction.

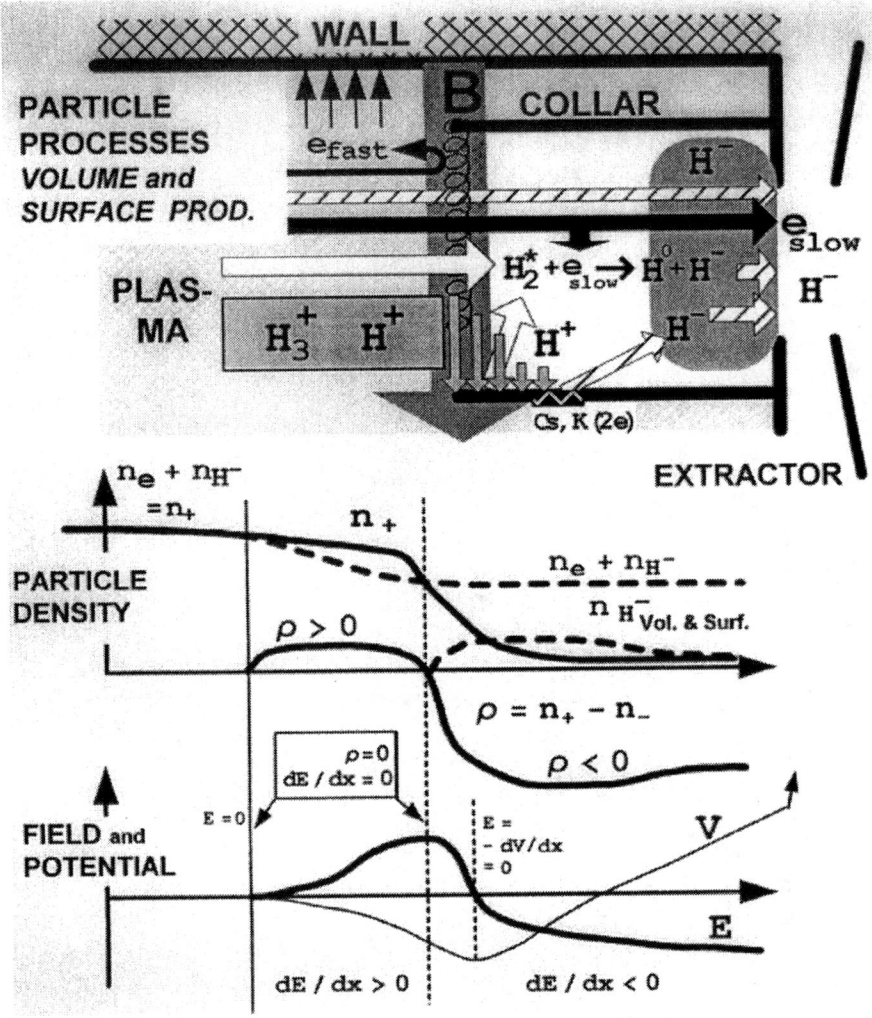

FIGURE 3. Particle processes in the collar region. Particle density and the resulting field and potential.

CONCLUSION

Based on Langmuir and collar measurements a new model for the plasma transition in H⁻ sources with collar has been developed. These measurements were possible due to the measurement friendly construction and the high reliability of the New HERA H⁻ RF Volume Source.

ACKNOWLEDGMENTS

The author is grateful for the contributions of the following colleagues at DESY : I.Hansen, H.Sahling and R.Subke. I wish to thank the technical groups at DESY for their support, and M. Lomperski of DESY for helpful suggestions to the wording of the report.

REFERENCES

1. Stockli, M. P., et al, *Rev. Sci. Instrum.*, **73** ,1007, (2002).
2. R. F. Welton et al, *Rev. Sci. Instrum.*, **73**, 1008-1012 (2002).
3. J.Peters, Proceedings of the 8th European Particle Accelerator Conference (June 2002).
4. J.Peters, Proceedings of the XX International Linear Accelerator Conference (August 2000).
5. J.Peters, Ph.D. thesis, Universität Frankfurt, 2001
6. R. Becker, K.N. Leung, W. Kunkel, *Rev.Sci. Instrum.* **69**, 1107-1109 (1998).
7. M. Leitner, D.C. Wutte, K.N. Leung, *Nucl. Instr. and Meth.* **A 427** 242 (1999).

Design, Operational Experiences and Beam Results Obtained with the SNS H⁻ Ion Source and LEBT at Berkeley Lab*

R. Keller,[#] R. Thomae,[#] M. Stockli,[@] and R. Welton[@]

[#] E. O. Lawrence Berkeley National Laboratory, Berkeley, CA 94720, USA

[@] Oak Ridge National Laboratory, 701 Scarboro Road, Oak Ridge, TN 37831, USA

Abstract. The ion source and Low-Energy Transport (LEBT) system that will provide H⁻ ion beams to the Spallation Neutron Source (SNS)** Front End and the accelerator chain have been developed into a mature unit that fully satisfies the operational requirements through the commissioning and early operating phases of SNS. Compared to the early R&D version, many features of the ion source have been improved, and reliable operation at 6% duty factor has been achieved producing beam currents in the 35-mA range and above. LEBT operation proved that the purely electrostatic focusing principle is well suited to inject the ion beam into the RFQ accelerator, including the steering and pre-chopping functions. This paper will discuss the latest design features of the ion source and LEBT, give performance data for the integrated system, and report on commissioning results obtained with the SNS RFQ and Medium-Energy Beam Transport (MEBT) system. Prospects for further improvements will be outlined in concluding remarks.

INTRODUCTION

Berkeley Lab has just completed building the linac injector (Front End, FE) for the Spallation Neutron Source project (SNS) and commissioning the entire system. The

---

* This work is supported by the Director, Office of Science, Office of Basic Energy Sciences, of the U.S. Department of Energy under Contract No. DE-AC03-76SF00098.

** The SNS project is being carried out as a collaboration of six US Laboratories: Argonne National Laboratory (ANL), Brookhaven National Laboratory (BNL), Thomas Jefferson National Accelerator Facility (TJNAF), Los Alamos National Laboratory (LANL), E. O. Lawrence Berkeley National Laboratory (LBNL), and Oak Ridge National Laboratory (ORNL). SNS is managed by UT-Battelle, LLC, under contract DE-AC05-00OR22725 for the U.S. Department of Energy.

main subsystems are the H⁻ ion-source, the low-energy beam-transport system (LEBT), the 2.5-MeV radio-frequency quadrupole (RFQ) accelerator, and the medium-energy beam-transport system (MEBT). Ion source and LEBT are the subject of this paper; their task is to create a 65-keV, 38-mA ion beam, to match and steer it into the RFQ, and to pre-chop it into mini-pulses of about 600 ns duration. The nominal duty factor is 6%, with 1-ms macro-pulse length and 60-Hz repetition rate.

Based upon the main design features of the SSC ion source [1], an R&D version of the SNS ion source was built first to demonstrate the viability of the chosen approach, utilizing an rf driven discharge inside a multicusp plasma generator with magnetic filter, cesium enhancement, and electron suppression at low energy [2]. This source version did not allow implementing cesium enhancement and electron suppression at the same time, but both features were proven to work satisfactorily in separate tests.

For the LEBT, a purely electrostatic focusing system was chosen, thereby avoiding time-dependent space charge compensation usually encountered with magnetic LEBTs. The basic design was derived from a successful proton LEBT that had been used to inject beam into an RFQ accelerator on a test bed at LBNL [3].

When the SNS project raised the performance goal for the Front End from 28 to 52 mA, the design of a second-generation source had already progressed, and it appeared justified to carry on with the construction of this so-called startup source that aimed at a beam current of 35 mA to be delivered through the LEBT. The LEBT attached to this source had been modified from the original design to accommodate up to 70 mA of beam current [4], and the highest measured current value was 40 mA [5].

The production version of the ion source and LEBT aims at generating and transporting a beam with 50-mA current, thought to be sufficient to satisfy the latest SNS design goal of 38 mA under the rather conservative assumption of 20% beam loss in the RFQ. As it turned out during the RFQ commissioning [6], this assumption might have been overly conservative because out of 36 mA, maximum, beam injected into the RFQ up to 33 mA were passed through, and negligible losses are anticipated for the MEBT. It is also quite probable that the quality of the LEBT beam will improve somewhat as less than the nominal 50 mA have to be generated by the ion source. By now, three plasma generators and one LEBT, as shown in Fig. 1, have been built and tested, and the specific design features and performance data of this production system are discussed in the following sections.

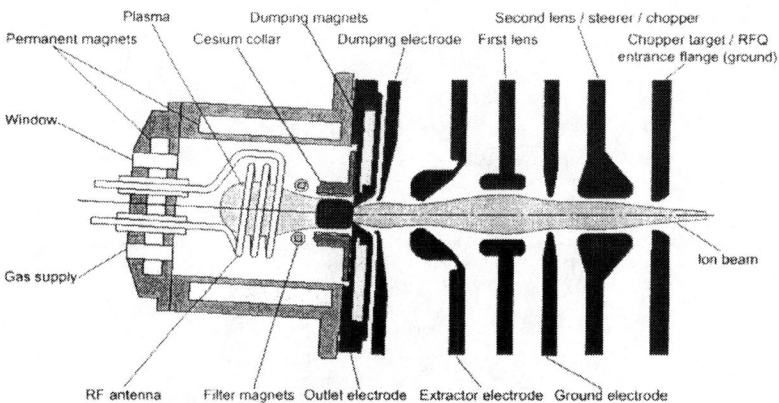

FIGURE 1. SNS Ion Source and LEBT layout, seen from the top. The actual orientation of filter and dumping magnet fields is orthogonal to the illustration plane. The width of the ion beam has been exaggerated in this figure to emphasize the focusing action of the two-lens LEBT. The second lens is split into four insulated quadrants to provide steering and chopping by applying separate pulsed and dc potentials.

ION SOURCE

The production-version ion sources aim at generating H⁻ beams of up to 50-mA current. Behind the LEBT, the goal for the normalized, transverse rms emittance of these beams is 0.2 π mm-mrad. For clarity, the term ion source in the context of this paper includes the plasma generator and electron dumping electrode, but no extractor which, instead, is considered to be part of the LEBT.

Plasma Generation

The design of the plasma generator is shown in Fig. 2. Tens of kW of 2-MHz rf power are sustaining the hydrogen discharge inside the multi-cusp vessel. The rf power is transmitted to the internal antenna through an inductive impedance-matching network. The antenna consists of a 2-1/2 winding copper coil, covered by a multi-layer porcelain coating [7]. An antenna with 0.25-mm coating underwent an endurance test at full duty factor, and the test was intentionally stopped after 107 hours and after verifying that the ion source delivered 20 mA of beam current at that time. An upgraded antenna with 0.8-mm thick 10-layer coating produces the same plasma density and

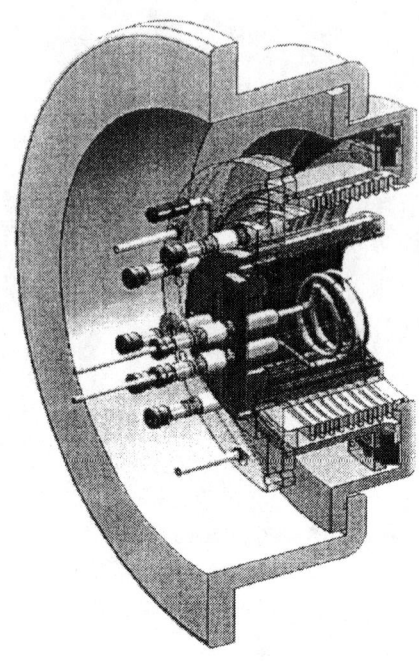

FIGURE 2. Multi-cusp plasma generator with rf antenna, inserted in the reentrant LEBT flange. The outlet plate and electron-dumping electrode are not shown here. The entire ion source can be pivoted horizontally around an axis in the outlet plane, on the right-hand side.

beam current for a given rf power level as the thin version (about 1.0 mA per kW of power) and is expected to last significantly longer.

A reliable method of plasma ignition at the beginning of every 1-ms pulse, was developed by maintaining a continuous, low-power discharge sustained by an additional 13.56-MHz rf system [8]. For reasons solely of practicality, we are using a capacitive impedance matcher for this continuous low-power rf. Initial difficulties were encountered with the 13.56-MHz amplifier switching off due to overload protection during the main pulses because of the impedance mismatch with the much denser plasma. This problem was eliminated by using an amplifier with higher nominal power rating than actually required to sustain the continuous low-density plasma alone.

During the development phase, we analyzed the light emitted from the plasma with a small spectroscope and were able to identify spectral emission lines of atomic hydrogen, cesium, and copper.[8] This technique will allow characterizing the ion source plasma on-line and can be extended to predict the end of the antenna life while beam production is still ongoing.

H⁻ Creation

Negative hydrogen ions are preferentially created in the space confined by the cesium collar, the magnetic filter field, and the outlet electrode, see Fig. 1. The filter field keeps energetic electrons that would destroy the H⁻ ions away from the collar region [9]. Volume production alone is sufficient to generate about 15 mA of beam current, but to reach the 50-mA level, cesium enhancement is needed. For that purpose, the collar is fitted with eight cesium-chromate containers and is thermally isolated from the source body to allow it heating up to several hundred degrees C. The presence of about 1/2 a mono-layer of cesium inner collar surface not only multiplies the abundance of negative ions in the discharge plasma by about a factor of three, but it also reduces the abundance of electrons in the extracted beam by one order of magnitude. Details of the enhancing cesium action are given elsewhere [10].

To best utilize the cesium, a freshly cleaned plasma generator is operated at full duty factor for about 15 min., heating the cesium collar to more than 500°C by forcing hot air through the collar wall and utilizing the rf power as a source of heat. After this initial conditioning, the collar is cooled down by room-temperature air and kept at about 280°C for optimal beam production. The cesium layer can then last for several days. Additional cesium reconditioning can be performed in-situ as needed. After controlled venting to atmospheric pressure, using dry nitrogen first, in order, for example, to change out a defective rf antenna, the source can be restarted without any cleaning because the amount of deposited cesium is very small.

Recently a new collar has been developed that seamlessly merges with the outlet aperture. This design brings three major advantages: 1), it provides for keeping the surfaces around the outlet aperture at the same temperature as the collar; 2), with the help of an isolating centering ring, it allows precise aligning of the collar to the axis of the outlet aperture and at the same time biasing the entire unit to an optimal potential with respect to the source body [11]; and 3), it allows modifying the contour of the outlet aperture as will be discussed below, without having to build another main outlet flange. This integrated collar/outlet aperture has been fabricated but so far not yet been tested in the SNS ion source. Its design is shown in Fig. 3.

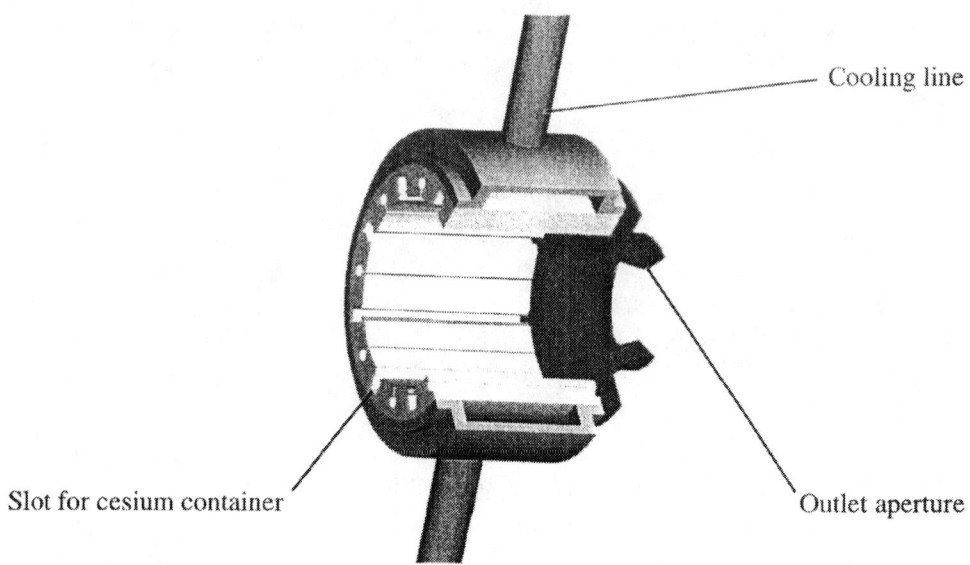

FIGURE 3. Integrated cesium collar with slots for cesium containers (left) and outlet aperture (right). Either heated or room-temperature air is conducted from the lower to the upper tube, passing around the collar, to keep it at optimal temperatures for initial cesiation or beam production.

LOW-ENERGY BEAM TRANSPORT

Electron Dumping

When a negative-ion beam is being extracted from a plasma, a substantial amount of electrons is extracted as well. These electrons create two kinds of problems: the first due to their space charge, especially near the outlet aperture where the particle velocities are low; and the second one due to the power load to the structure where the electron beam is deposited. Unfortunately, trying to solve the second of these problems by separating the electrons at low energy from the ions increases the severity of the first problem. The compromise that was sought with the design of the SNS ion source consists in depositing the electrons on a dedicated 'dumping electrode' at moderate energies of about 5 keV as compared to the full ion-beam energy of 65 keV. Mechanically, this dumping electrode can be considered to be a part of the ion source proper because it is supported by and insulated against the outlet electrode. A spark gap ensures that no excessive voltage can be built up between these two electrodes when the main extraction voltage breaks down.

The separation process is accomplished by a set of permanent magnets inserted inside the outlet electrode in a so-called Halbach configuration [12]; the magnets create a sharply peaked dipole field in the extraction gap. Because of the uncertainty of the actual plasma meniscus location with respect to the dumping field, aggravated by the fact that this location might well depend on the plasma density inside the collar volume as well as on the degree of cesiation, the actual electron trajectories are not well enough known under all conditions, and in most cases some fraction of the electrons misses the dumping electrode entirely. These electrons are accelerated to full beam energy and hit either the extractor electrode or its support structure. A water-cooled shield has now been installed on the support structure to take care of the heat load. It is expected that the new, integrated collar/outlet electrode described above will definitively resolve this problem.

Because of the steering action of the dumping magnetic field, not only on the electrons, but also on the ions, the entire ion source is tilted by an adjustable angle of about 3° with respect to the LEBT axis.

Beam Formation

When the SNS Ion Source and LEBT were developed, only positive-ion codes were available to us for simulation work. The code IGUN [13] had proven to be very reliable in predicting trajectories of positive ions, and we used it by implicitly assuming the negative ions to be protons and adding a fixed amount of current to the expected ion beam current to represent the space charge of the electrons in the zone between outlet and dumping electrodes [4]. Even though this method was good enough to get a very useful LEBT design, these simulations proved to be inadequate for the purpose of optimizing the extraction system details. In the meantime, the code PBGUNS [14] had been extended by its author to properly address negative-ion extraction problems, and with its help an improved outlet-electrode design was created that was implemented in the new, integrated collar/outlet electrode.

The main difference between IGUN and PBGUNS results is their prediction where the plasma meniscus would be anchored on the outlet aperture, as illustrated in Fig. 4. Experimental results discussed below support the validity of the PBGUNS results. The non-uniformly curved plasma meniscus as predicted by these latter simulations for the original SNS ion-source outlet geometry produces significantly distorted emittances.

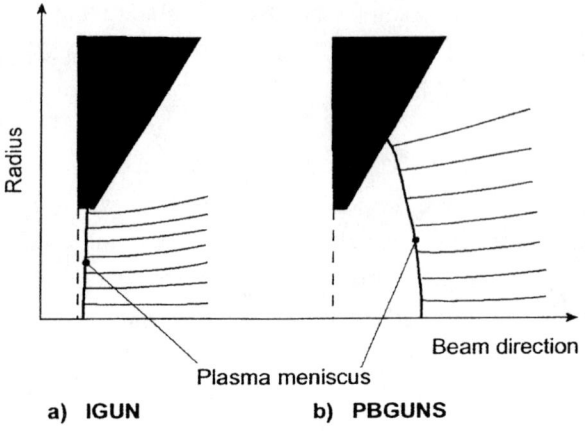

FIGURE 4. Schematic simulation results of the beam formation process with the original SNS ion-source outlet geometry. In close agreement with actual emittance measurements, the PBGUNS simulations (b), suggest that the plasma meniscus is not anchored at the edge of the outlet aperture but protrudes significantly beyond the ideal outlet plane. This effect leads to a larger size of the meniscus than was predicted for protons by IGUN and to a non-uniform meniscus curvature, both resulting in larger and more distorted emittances.

A systematic study of outlet-aperture contours [15] indicated that a longer channel such as the one shown in Fig. 3, as opposed to the knife-edge shape seen in Fig. 4, would lead to a smaller and essentially flat meniscus, resulting in significantly smaller emittance size.

Principal LEBT Functions

Apart from the beam formation, the main purpose of the LEBT is to transport the beam to the RFQ and give it Twiss parameter values matching the injection requirements. Efficient pumping of the gas load produced by the plasma generator is another task that was addressed by giving the electrode support structures highly transparent shapes, as shown in Fig. 5.

The focusing action of the electrostatic two-lens system, captured in a tuning matrix, works as predicted by simulations, but generation of less than nominal beam current results in a narrower beam size inside the first lens and effectively attenuates its focusing power. For that reason, the extraction gap was widened by 4 mm as compared to the nominal size of 20 mm to handle beams in the 35-mA range during the

FIGURE 5. View of the LEBT electrodes inside the vacuum tank, from the downstream end. The first electrode seen is the center electrode of the second lens, split into four isolated quadrants.

RFQ commissioning. The widened gap allows the beam to expand more and thus enables the first lens to carry its share of the focusing action.

The functions of pre-chopping and static steering are implemented by applying pulsed voltage signals and, independently, dc potentials to the four quadrants of the center electrode of the second lens. Chopping voltages of ±2.5 kV amplitude are applied to adjacent pairs of quadrants such as to deflect the beam alternatingly into the four diagonal directions and thus distribute the local heat load on the ring-shaped chopper target that is imbedded in the RFQ entrance flange. Not all of the chopped beam is intercepted by this ring target; the remaining particles are deposited inside the RFQ cavities, well away from the vane tips that are oriented in the horizontal and vertical directions.

BEAM RESULTS

The nominal beam-current goal of 50 mA pulse average measured at full 6% duty factor downstream of the LEBT was reached about a year ago, still with the nominal gap width installed. The Faraday cup used for all these measurements is magnetically isolated and electrically biased to suppress secondary electrons as well as electrons that might have been transported together with the negative ions. The peak current at the beginning of every pulse even reached 68 mA, as shown in Fig. 6. The pulse shape had not been optimized for uniformity, and if the initial peak cannot be suppressed by just applying suitable discharge conditions the option of shaping the 2-MHz rf-power waveform is available. With the extraction gap between dumping electrode and LEBT increased by 4 mm for RFQ commissioning, as discussed above, up to 36 mA were measured, and the pulse shape was much more uniform.

The horizontal and vertical emittances of a focused 33-mA LEBT beam are shown in Fig. 7. They were measured using an Allison-type emittance device with electrostatic angle scan [16]. Pronounced distortions are evident slightly beyond the 10% intensity level, but after subtracting background signals from the raw data, the rms sizes are very close to or even better than the nominal values.

A round-the-clock endurance test of the ion source and LEBT was conducted over more than a week, continuously producing beam with about 25 mA current at 3% duty factor. This test led to the elimination of several minor technical flaws and was very successful overall. The sparking rate steadily improved to about once per hour, and operations were interrupted only a few times, mostly because of external events that affected the ion source operation. The test proved that the beam-generating system is ready to support commissioning of the SNS accelerators and even SNS operations for the first few years, and in addition it helped dealing with a few remaining design weaknesses.

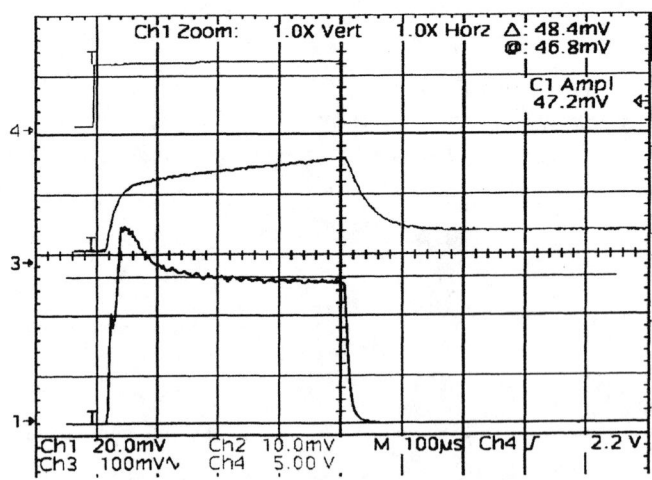

FIGURE 6. Beam pulses (50 mA average, 68 mA peak) measured downstream of the LEBT with the nominal extraction gap.

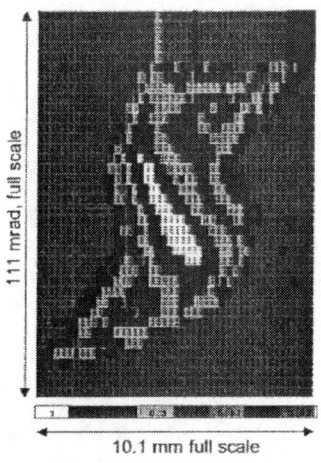 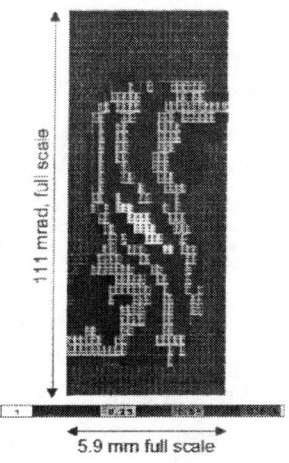

Horizontal, 0.22 pi mm mrad   **Vertical, 0.15 pi mm mrad**

FIGURE 7. Horizontal and vertical LEBT emittances for a 33-mA beam out of the LEBT, after subtraction of background signals: 2.1% of peak intensity horizontal and 1.3% vertical, respectively.

In fact, the commissioning of the SNS RFQ and MEBT at low duty factor [6] was aided by very stable performance of Ion Source and LEBT. On the last day of integrated testing, a beam current of 50 mA was transported through the MEBT, exceeding the project goal by nearly 30%. Emittances of diverging beams measured 450 mm downstream of the RFQ do not show any signs of residual distortions; apparently the RFQ clips the most aberrant parts of the injected phase-space distributions while still allowing more than 90% of the beam to pass through.

The functionality of the LEBT pre-chopper system has been tested as well, and rise and fall times of 25 ns were measured, beating the nominal requirement by a factor of 2 and reducing the thermal load on the MEBT chopper-target in proportion. The beam signals were not clear enough to allow a precise determination of the pre-chopper attenuation factor, but a value around 1×10^{-3} appears quite plausible from extrapolations of results obtained at less than nominal chopping voltages.

OUTLOOK

The Ion Source and LEBT systems built for the SNS Linac injector have come a long way towards fulfilling their nominal performance goals in terms of beam current and quality as well as overall functionality. The RFQ commissioning results even appear to allow for some margin in terms of the rms emittance sizes. It remains to be seen, however, how much the noted aberration tails of the LEBT beam are going to contribute to halo generation in the MEBT and even the Linac sections farther downstream. The rms emittance-concept evidently is too coarse a measure to ascertain such effects. Even though the SNS project is going to provide for halo scrapers in the MEBT a reduction of the emittance distortions after the LEBT appears to be desirable, and the beam formation system offers itself for the first steps in that direction.

Quite some hope rests on the expected performance of the integrated collar/ outlet unit with improved outlet contour, but if this turned out to be insufficient one could go further along this path by implementing an electron dumping system at intermediate energies around 15 keV as used already elsewhere [17]. The advantage of such an approach lies in the fact that the parasitic electrons are at first being extracted in a geometry with cylindrical symmetry, avoiding the build-up of asymmetric space-charge forces that could act on the very slow ions near the meniscus. When the electrons are finally being separated from the ion beam their own speed is high enough that asymmetric space-charge forces become essentially negligible. The price for utilizing such a dumping system is to be paid in having to remove a higher thermal load from the catcher electrode, and possibly in a higher rate of x-ray emission.

It is debatable if one should work towards the elimination of the epoxy-based joints in the present LEBT assembly. Of course, overheating of these structures could lead to ungluing and downtime to execute the necessary repairs, but uncontrolled overheating of the LEBT electrodes for lack of a 'mechanical fuse' could lead to even more serious deformation of these parts and even worse loss of time. Installing water-cooling lines as already initiated with the present extractor support appears to be the better remedy in the long run, in addition to carefully avoiding unwanted heat deposition in the first place.

The actual lifetime of an ion source is a never-ending topic for accelerator projects, and for the SNS Ion Source, the rf antenna appears to be the most critical element in this respect. Very significant progress has been made with the antenna development reported in this paper, but the ultimate lifetime of the currently employed antennas under nominal operation conditions is simply not yet known. While this issue might be relevant for SNS only several years from now, when the actual beam-production phase of the facility has started, it certainly would be good to know much earlier that the nominal time of three weeks between scheduled services can be reached at will.

To solidly achieve the stated goals, a new approach to the discharge technology might well be required, and two options are currently being considered among the SNS collaboration. They involve either use of an external antenna such as the one that gave very respectable results in low duty-factor operation [18], or, as a more fundamental change, a pulsed-dc discharge mode supported by a plasma cathode.

ACKNOWLEDGEMENTS

The authors would like to acknowledge the support by a large number of SNS staff at LBNL and ORNL, not necessarily listed in the references, who supported this work. Thanks are due in particular to R. Gough, R. Yourd, R. DiGennaro, A. Ratti, S. Lewis, T. Schenkel, D. Cheng, K. N. Leung, J. Greer, J. W. Staples, D. Syversrud, W. Abraham, T. Kuneli, N. Ybarrolaza, C. Lionberger, P. Cull, M. Hoff, J. Pruyn, R. MacGill, M. Monroy, M. Regis, J. Dougherty, D. Garfield, and K. Barat.

REFERENCES

1. Saadatmand, K., Arbique, G., Hebert, J., Valicenti, R., and Leung, K. N., "Performance of a High Current H⁻ Radio Frequency Volume Ion source," *Rev. Sci. Instr.* **67** (3), p. 1318-1320 (1996).
2. Leitner, M. A., Cheng, D. W., Mukherjee, S. K., Greer, J., Scott, P. K., Williams, M. D., Leung, K. N., Keller, R., and Gough, R. A., "High-Current, High-Duty-Factor Experiments with the RF Driven H- Ion Source for the Spallation Neutron Source," *PAC '99*, New York (1999).
3. Staples, J. W., Hoff, M. D., and Chan, C. F., "All-electrostatic Split LEBT Test Results," *Linac '96*, (1996).
4. Reijonen, R. J., Thomae, R., and Keller, R., "Evolution of the LEBT Layout for SNS," *Linac 2000*, Paper MOD19, Monterey (2000).
5. Thomae, R., Bach, P., Gough, R., Greer, J., Keller, R., and Leung, K. N., "Measurements on the H - Ion Source and Low Energy Beam Transport Section for the SNS Front-End Systems," *Linac 2000*, Paper MOD09, Monterey (2000).
6. Keller, R. for the SNS Front-End Systems team, "Commissioning of the SNS Front-End Systems at Berkeley Lab," submitted to EPAC '02, Paris (2002).
7. Welton, R. F., Stockli, M. P., Kang, Y., Janney, M., Keller, R., Schenkel, T., Thomae, R., and Shukla, S., "Ion Source Antenna Development for the Spallation Neutron Source," *Rev. Sci. Instrum..* **73** (2), p. 1008 - 1012 (2002).
8. Schenkel, T., Staples, J. W., Thomae, R. W., Reijonen, J., Gough, R. A., Leung, K. N., Keller, R., Welton, R., and Stockli, M. P., "Plasma Ignition schemes for the Spallation Neutron Source Radio-Frequency Driven H⁻ Source," *Rev. Sci. Instrum.* **73** (2), p. 1017 - 1019 (2002).
9. Leung, K. N., "Negative Ion Sources," in *"The Physics and Technology of Ion Sources,"* edited by Brown, I. G., John Wiley & Sons, New York, p. 355, (1989).
10. Welton, R., Stockli, M., Keller, R., and Thomae, R., "Enhancing Surface Ionization and Beam Formation in Volume-type H- Ion Sources," submitted to *EPAC '02*, Paris (2002).
11. Peters, J., "The Plasma-Vacuum Transition in RF Sources for Negative Hydrogen Ions," *Rev. Sci. Instrum.* **73** (2), p. 900 - 902 (2002).
12. Halbach, K., *Nucl. Instrum. Methods* **169**, p.1 (1980).

13. Becker, R., "New Features in the Simulation of Ion Extraction with IGUN," *EPAC '98*, Stockholm (1998).

14. Boers, J. E., *PBGUNS Manual*, available through Thunderbird Simulations, Garland, TX, 75042.

15. Welton, R. F., Stockli, M. P., Boers, J. E., Rauniyar, R., Keller, R., Staples, J. W., and Thomae, R. W., "Simulation of the Ion Source Extraction and Low-Energy Beam Transport Systems for the Spallation Neutron Source," *Rev. Sci. Instrum.* **73** (2), p. 1013 - 1016 (2002).

16. Allison, P. W., Sherman, J. D., and Holtkamp, D. B., *IEEE Trans. Nucl. Sci.* **NS-30**, 2204 (1983).

17. Oguri, H., Okumura, Y., and Hasegawa, K., "Development of an H- Ion Source for the High Intensity Proton Linac," *Rev. Sci. Instrum.* **73** (2), p. 1021 - 1023 (2002).

18. Peters, J., "Internal versus External RF Coupling into a Volume Source," *EPAC '02*, Paris (2002).

Status of an RF Negative Hydrogen Ion Source using Transformer Coupled Plasma Source

I.S. Hong, H.D. Jung, Y.S. Park, and Y.S. Hwang

*Dept. of Nuclear Engineering, Seoul National University,
Seoul 151-740, KOREA*

Abstract. A negative hydrogen ion source based on the transformer coupled plasma (TCP) source has been developed as a long-lifetime continuous power (CW) ion source. H-minus beam currents of 3.3mA without cesium injection at the extraction voltage of 20kV are extracted from high-density radio frequency (RF) plasmas generated by 13.56MHz, 1.5kW RF power. Extracted beam currents have been measured at various distances from the ion source using a moveable Faraday cup. Effects of transverse magnetic fields from filter magnets are discussed.

INTRODUCTION

High-current, negative ion sources have been developed with RF plasmas for accelerators such as the national spallation neutron source (NSNS) [1]. Although RF antennae can promise longer lifetime than filaments in magnetic bucket ion sources, lifetime of an RF antenna is still limited since the antenna is immersed in the plasma. Various insulation materials are attempted to lengthen antenna lifetime further [2]. More successfully, an insulating cylinder has been inserted in front of the RF antenna to prevent the antenna from direct contact to plasmas[3]. As an advanced concept, a negative ion source based on a transformer coupled plasma (TCP) source which has the RF antenna outside of the plasma vacuum chamber has been developed as a high-current continuous power(cw) negative ion source[4].

In this paper, the present status of the ion source based on TCP will be reported. The experimental setup for the ion source is briefly described in section II. Section III presents characteristics of the H-minus ion source. In addition to overall performance of the source, effects of transverse magnetic fields applied in front of the extraction system are discussed in detail. In the last section, results will be summarized and concluded.

EXPERIMENTAL SETUP

An overall experimental setup for the TCP negative ion source is shown in Fig. 1. The plasma generation chamber is made of stainless steel in double-wall configuration to provide sufficient water-cooling for high power-density plasma operations. An RF antenna for plasma generation is located opposite to the extraction system and outside of the chamber isolated with either quartz or alumina plates as shown in Fig. 1. In the presence of sufficient insulation of up to 40kV between the RF antenna and plasmas, the RF power supply and matching unit are maintained at the ground level instead of high-voltage level, which makes the ion source operation much easier and more safe. Multi-cusp magnetic configuration around the plasma chamber is implemented to confine high-density plasmas that are generated with 13.56MHz, 2kW RF power.

A single-hole, two-electrode extraction system has been constructed to extract the ion beam. The diameter of the extraction hole and the extraction gap distance between two electrodes are 4mm and 9mm, respectively. High voltages of up to 30 kV have been applied to extract beams at the extractor. In front of the extraction electrode, a set of permanent magnets is placed to filter out fast electrons prior to H-minus beam extraction. A separate water-cooling path is provided to make sure filter magnets are sufficiently cooled even for high-density plasma operations. To get sufficient H-/e ratio, the strength of the filtering field has been adjusted. Effects of transverse magnetic fields from this magnetic filter system are discussed in more detail in the next section.

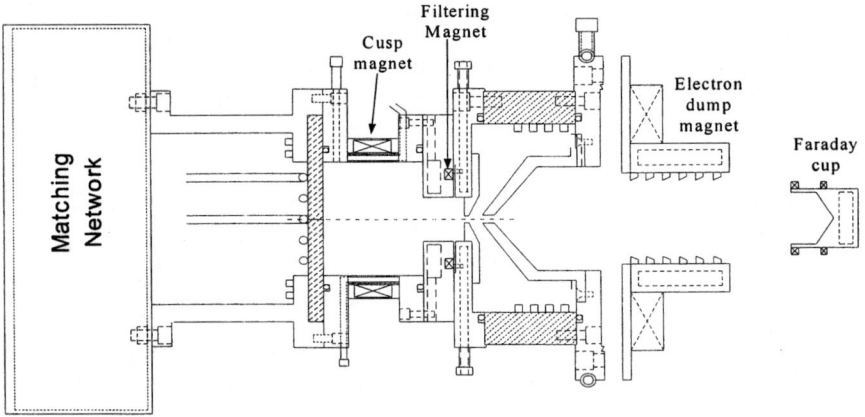

FIGURE 1. An overall experimental setup of the TCP ion source.

To measure H-minus beam currents and beam divergence accurately, a Faraday cup has been installed and moved along the beam extraction axis. For high current and high extraction voltage operations, the cup is cooled with water and attached with permanent magnets near the entrance of it to reduce escaping secondary electrons generated by high-energy negative ion bombardments. Another set of permanent magnets has been installed as a bending magnet in front of the Faraday cup to dump electron beams out of the main H-minus beam.

ION SOURCE CHARACTERISTICS

High-density plasmas are generated with the TCP source for hydrogen gases. With the help of a magnetic multicusp field, the threshold power for high-density mode operations can be reduced down to 1.3kW even at a lower gas pressure with the gas-feeding rate of 25sccm. After increasing the number of antenna turns from two to three-and-half, the threshold power for the hydrogen discharge was reduced further to below 1kW. With the three-and-half turn antenna, high-density plasmas with hydrogen gases can be obtained at the power level of above 1kW. With a simple, single-hole, two-electrode extraction system, electron and H-minus beam currents are extracted.

When both gas pressures and extraction voltages are fixed, the electron beam current increases with the applied RF power as shown in Fig. 2, indicating a linear dependency of total extracted beam currents on applied RF power, i.e. plasma densities. Effects of beam divergence can be estimated from the measured negative hydrogen beam currents by varying the position of Faraday cups. As the Faraday cup moves away from the ion source, collected negative beam currents decrease and H-/e ratios get lower, indicating diverging H-minus beams.

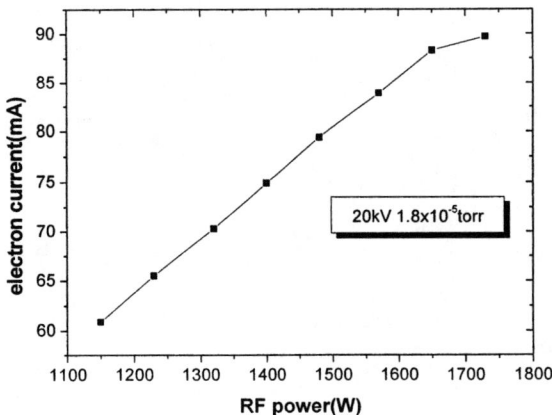

FIGURE 2. RF power dependency of extracted electron beam currents.

At the fixed gas-feeding rate and the fixed RF power, i.e. with the same plasma density, H-/e ratios are increased as extraction voltages are increased as shown in Fig. 3(a), indicating the extraction voltages are in the range of under-focused regimes, i.e convex meniscus. With the given plasma density, an optimum extraction voltage will be determined when under-focused beams become over-focused beams as the extraction voltages are increased. Since shapes of ion emissive surfaces are determined by the relative strengths between the plasma density and the extraction voltage, higher extraction voltages will be required for the optimal beam focusing as plasma densities increase with higher RF power as shown in Fig. 3(b).

An important aspect to increase negative hydrogen beam currents is increasing the ratio of H-/e. The ratio can be increased with low electron temperatures by eliminating fast electrons in front of the plasma electrode, which can be achieved by providing an optimum transverse magnetic field with filter magnets. As the transverse magnetic field strength of filter magnets has been varied from 40 Gauss to 250 Gauss, characteristics of the extracted beams are shown in Fig. 4. As the magnetic field becomes stronger, electron beam currents drops significantly due to fast electron losses as expected. At low magnetic fields below 100 Gausses, relatively low e/H- ratios of about 20 as well as high electron beam currents have been obtained resulting in highest H-minus beam currents. Even with such a low magnetic field of 40 Gauss, electron temperatures near the extraction region are maintained low enough to achieve such a high H-/e ratio. Such a low electron temperature at the extraction region may be expected from the fact that RF heating power is localized at the opposite side of the extraction region in the TCP ion source. Highest H-minus beam current density of 24.6mA/cm^2 at the extraction voltage of 22.5kV has been obtained at this low magnetic field of 40 Gausses. However, H-minus beam currents cannot be increased further at this high current density operation since operations with higher extraction voltages are hindered at the power limits of about 1.5kW. On the other hand, operations at low extraction voltages with high-currents are limited by arcing problems at such a high plasma density that plasma emitting boundaries bulge out too much at relatively low extraction electric fields.

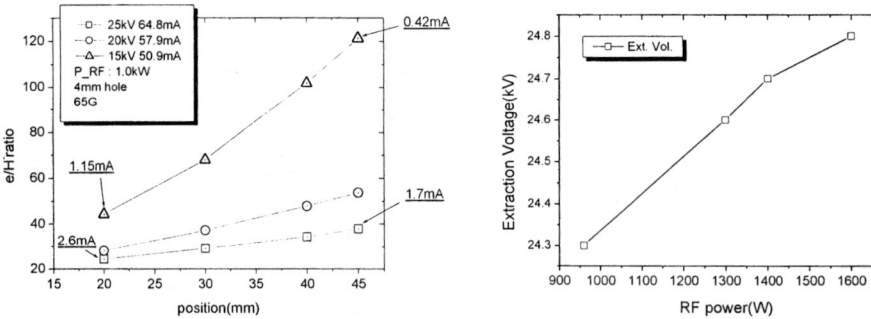

FIGURE 3. (a) e/H- ratios measured at different distances from the ion source for various extraction voltages (b) Extraction voltages for minimal beam divergence as a function of RF power.

Highest H-minus beam currents of 3.3mA can be obtained at low extraction voltage of 20kV by operating at slightly lower plasma densities with higher transverse magnetic field of 65 Gauss. Another power supply operating at high currents up to 120mA with the maximum voltage of 20kV has been used to get this high H-minus current with larger extraction hole diameter of 5mm at the beam current density of 16.8mA/cm^2. For further increase of negative hydrogen currents, either extraction power supplies with higher power or a three-electrode extraction system are required.

Another interesting characteristic has been observed with stronger transverse magnetic field from the results shown in Fig. 4. Plasma emitting boundaries are determined from the relative strengths of plasma densities and applied electric fields. For the given electric field generated by the extraction system, extracted beams change from over-focused to under-focused according to the shape of plasma sheaths as plasma densities become higher. Although plasma densities are reduced significantly with high magnetic field of 250 Gauss, transitions from under-focused beam to over-focused beam occur at much higher extraction voltages than those of high plasma densities at low magnetic field do. This may indicate that emitting boundaries of plasmas are bulging out at the plasma electrode even at the low plasma density for the given extraction field because the strong magnetic fields from the filter magnets determine the shapes of plasma emitting boundaries, i.e. plasma sheaths. Based on this, the emitting boundary of plasmas may be controlled at the plasma electrode by applying either concave or convex shapes of magnetic fields. With the concave shape of the magnetic field, emitting boundaries can be maintained to be concave even with higher plasma densities at the given extraction voltage, resulting in high extraction currents without the plasma bulging out and without resultant arcing.

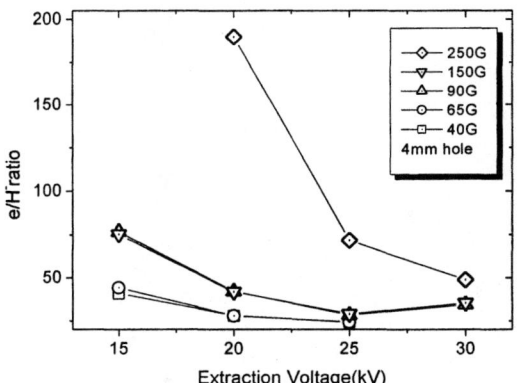

FIGURE 4. e/H- ratios of the extraction voltage for various transverse magnetic field strengths.

CONCLUSION

Negative hydrogen beam currents of up to 3.3mA at the extraction voltage of 20kV are extracted with the RF power of 1.5kW without any additives such as Cesium and Argon. The e/H- ratio of about 20 has been obtained with relatively low filtering fields of less than 100 Gauss, which may be due to intrinsically low electron temperatures near extraction region of the TCP ion source. An extraction system with three electrodes is required and under development for higher H-minus beam currents of 10mA with the beam energy of 30kV without cesium injection in continuous operation mode. In addition, controlling the emitting boundary of plasmas for better ion beam characteristics may be possible by applying appropriate magnetic field structures near the plasma electrode.

ACKNOWLEDGMENTS

This work supported by both KISTEP and BK21.

REFERENCES

1. Leung, Ka-Ngo, *Rev. Sci. Instrum.* **71**, 1064-1068 (2000).
2. Lee, Y., et. al., *Rev. Sci. Instrum.* **69**, 1023-1025 (1998).
3. Peters, J., *Rev. Sci. Instrum.* **69**, 992-994 (1998).
4. Hong, I.S., Hwang, Y.S., and Cho, Y.S., *Rev. Sci. Instrum.* **73**, 979-982 (2002).

A high-intensity H⁻ ion source

K. Volk, H. Klein, A. Maaser, U. Ratzinger

Institut für Angewandte Physik, Universität Frankfurt, 60054 Frankfurt, Germany

Abstract. At the University of Frankfurt, a high current H⁻ source has been developed and tested. The ion source is of the volume type with four tungsten filaments. It consists of a cesium seeded multicusp plasma generator equipped with a variable filter magnet. Due to improvements of the cesium injection method, the beam current density has been enhanced up to 153 mA/cm^2. Thus, an H⁻ beam current of 120 mA @ 35 kV has been extracted using an emission opening radius of 5 mm. For a limited time 140 mA H⁻ could be extracted.

INTRODUCTION

The development of pulsed high brightness H⁻ ion sources for RFQ injection is presently a strong interest. The normalized rms beam emittance in front of the RFQ should be 0.1π mm mrad or smaller and the beam noise level should be below +/- 1%. These requirements have to be fulfilled with great constancy and reliability. The source must operate over an acceptable period of time.

The Institut für Angewandte Physik of the Frankfurt University is concerned with the development of such an H⁻ ion source. As a promising candidate for this task, the so-called HIEFS was chosen. It was developed in Frankfurt [1] and belongs to the volume type family.

EXPERIMENTAL SETUP

A schematic cross-sectional view of the experimental setup is shown in Figure 1. The plasma chamber of the ion source is made of a water-cooled copper cylinder. It is surrounded by 10 CoSm magnets in cusp field arrangement. Near the chamber axis, four tungsten filaments (1.8 mm diameter) are mounted. The front end of the chamber is enclosed by the plasma electrode. It is electrically connected with the anode. An electromagnet is installed in the flange of the plasma electrode. Its transverse magnetic field (B_f) acts as an electron filter.

The arc power is provided by an array of capacitors (C_{tot} = 0.53 F) and gated by a high current switch. The pulse generator used allows arc powers up to 50 kW during 0.15 to 1.2 ms pulses and variable repetition rates from 1 to 400 Hz. Furthermore, for beam currents beyond the current limit of our extraction power supply (65 kV/300 mA), its charge also has to be accumulated in a capacitor (C = 3.5 µF). A wiring diagram of the ion source is presented in Figure 2.

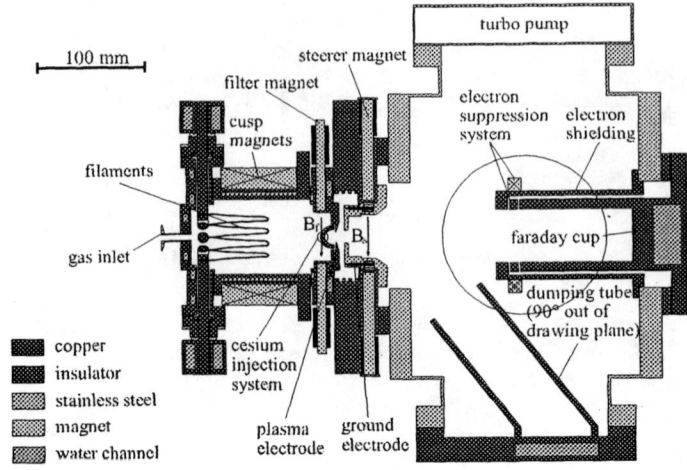

FIGURE 1. Schematic drawing of the experimental setup.

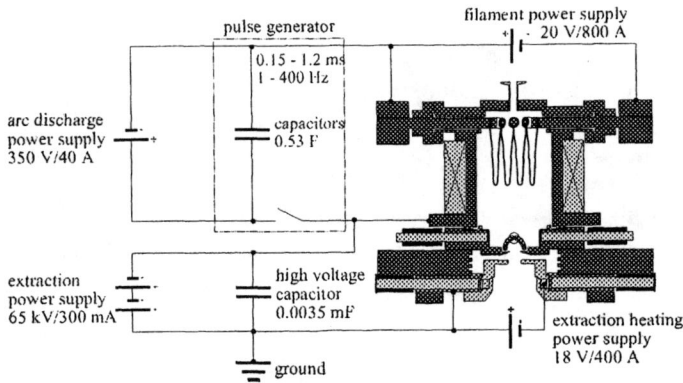

FIGURE 2. Wiring diagram of the H⁻ ion source.

An external oven is mounted on the flange of the plasma electrode for introducing cesium. By means of two small pipes, the cesium vapor is deposited close to the outlet aperture. The whole system is temperature controlled. In order to extract the H⁻ ions, a single hole diode extraction system with an aspect ratio of 0.8 is used. A heater is integrated in the ground electrode to run the extractor at temperatures around 250 °C. In this way, a cesium deposition on the ground electrode is avoided which would lead to high voltage breakdowns in the extraction gap otherwise.

As the extraction of H⁻ ions is always accompanied by the extraction of electrons, a device for electron beam removal is necessary. Our concept is to dump the electron beam behind the extractor at full beam energy in a water-cooled cup. Due to the extension of the filter magnet field into the gap of the extractor, the electron beam is deflected out of the beam axis. As the deflection angle of the electron beam depends on the filter field strength and the extraction energy, an additional so-called "steerer magnet" is necessary (B_s) to contrive the electron beam in the dumping tube.

The ion beam diagnostic consists of a water-cooled Faraday cup, encapsulated in a grounded screen. It is equipped with electrostatic and magnetic electron shielding.

EXPERIMENTAL RESULTS

There are two dominant processes for H⁻ generation in this ion source: volume and surface production. Both processes can be intensified by supplying cesium to the plasma chamber [2]. In operation with cesium, the H⁻ emission current is up to 4.5 times higher and the e/H⁻ ratio is about 7 times lower compared to operation without cesium.

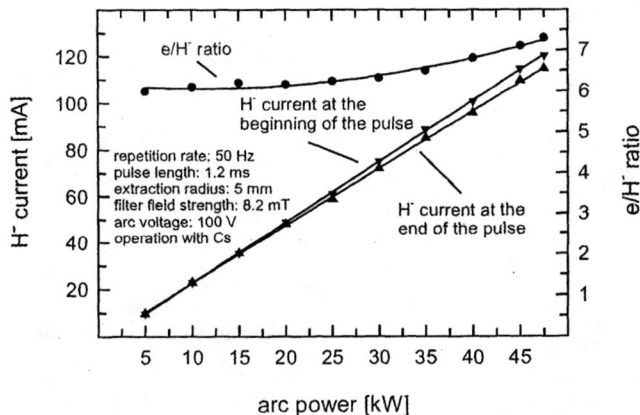

FIGURE 3. H⁻ current and e/H⁻ ratio vs. the arc power.

Figure 3 illustrates the H⁻ current as a function of the arc power for cesiated source operation. The graph is split for higher arc powers: the upper graph shows the H⁻ current at the beginning of the pulse, the lower graph indicates the H⁻ current at the end of the pulse. This splitting, or current decrease during the pulse, is due to the "burn in" of the arc discharge. Especially for higher arc powers, it takes more time until the discharge achieves its balance. For an arc power of 47.5 kW, the ion source produces 120 mA H⁻ at the beginning of the pulse and 115 mA H⁻ at the end of the pulse. The former value corresponds to an emission current density of 153 mA/cm². In this mode of operation, the required extraction voltage for a matched beam is 33 kV (E_{gap} = 5.3 kV/mm). It is remarkable that the H⁻ current increases linearly as a function of the arc power. Hence, a further enhancement of the H⁻ current is expected for arc powers beyond 50 kW. For short times 140 mA have already been achieved.

Figure 3 also shows the e/H⁻ ratio as a function of the arc power. The plot displays a small growth rate. For full arc power the e/H⁻ ratio is about 7. This value corresponds to an electron current of 850 mA, equivalent to a mean electron beam power of 1.7 kW. Almost 100 % of the electron beam is captured in the dumping tube where the deposited energy of the electron beam is removed easily.

Figure 4 presents the evolution of the H⁻ current during a 1.2 ms pulse. After a rise time of 75 μs the curve reaches its maximum, shows a reduction of 4 %, and falls down within 35 μs. For an arc power of 47.5 kW and a matched beam the noise level is less than 1.5 % (peak to peak). To attain such a low level, both the H⁻ and the electron beam must not hit any electrodes. An ion source operation without cesium leads to a considerably faster current decrease during the pulse. Also, the beam noise level is dramatically higher.

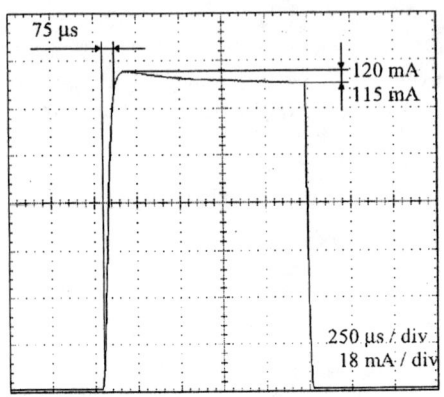

FIGURE 4. Oscillogram of the H⁻ beam.

In a further experiment, the influence of the filter field strength on the H⁻ current and the e/H⁻ ratio was investigated. The measurements were made at an arc power of 35 kW. As illustrated in Figure 5, raising the filter field strength reduces the e/H⁻ ratio while having little influence on the H⁻ current. Consequently, the ion source is operated with a filter field strength of 8.2 mT.

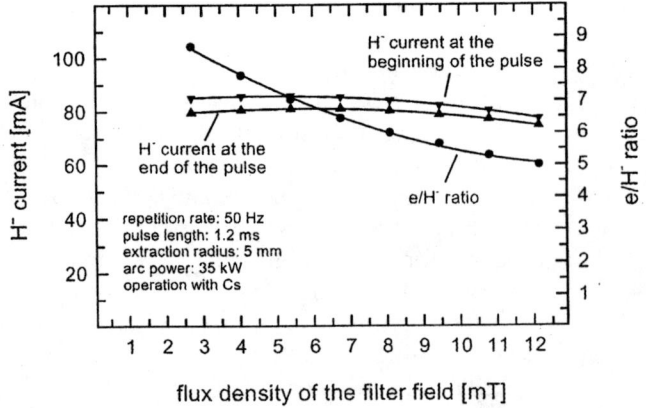

FIGURE 5. H⁻ current and e/H⁻ ratio vs. filter field strength.

In the course of the experiments, the dependence of the H⁻ current and the e/H⁻ ratio on the plasma electrode temperature was investigated. As depicted in Figure 6, an increase of the temperature yields higher H⁻ currents. For about 200 °C, the H⁻ current is about 1.3 times larger compared to its value at room temperature. Over the same temperature range, the e/H⁻ ratio drops by a factor of 2.

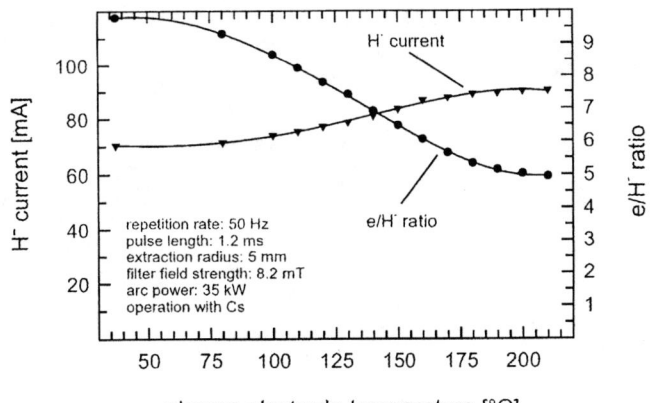

FIGURE 6. H⁻ current and e/H⁻ ratio vs. plasma electrode temperature.

Besides a high H⁻ current, an appropriate ion source lifetime is another important requirement. For a given operational gas, the lifetime is limited by the wear of the filament, which is approximately antiproportional to the duty cycle and the arc power. In order to obtain a long ion source lifetime, one should employ several filaments with large cross sections. This of course requires a filament power supply of the proper size. In other words, many thick filaments - which unfortunately have to be operated with a large power supply - are favorable for a long lifetime of the ion source. With the configuration described in this paper, the ion source was operated for 190 h at an arc power of 40 kW. After this time, the experiment was interrupted to check the condition of the filaments. Their diameter was reduced from 1.8 mm at the beginning to 1.5 mm. Since the filament can be used down to a thickness of about 0.9 mm, the ion source lifetime should be about 14 days.

As cesium is injected on demand only, in the normal mode of operation the injection system is off. This means there is also a wear of the cesium layer which has to be rebuilt every 10 to 18 h. To achieve a controlled cesium deposition, a specific temperature profile of the plasma generator has to be kept. To account for this fact, the arc power has to be reduced for about 15 min.

So far, the beam emittance has been not measured with our slit-grid emittance measurement device. Nevertheless, the H⁻ beam was recorded with a video camera to get a rough emittance estimation. After an optical analysis, the beam radius and the divergence angle of the 120 mA - 33 keV H⁻ beam could be estimated to 3 mm and 10 mrad respectively. Hence, a phase space area emittance of 30π mm mrad is covered. For a KV distributed beam, the rms emittance is 7.5π mm mrad and the corresponding normalized rms emittance is 0.07π mm mrad.

CONCLUSIONS

In the present article, the status of the Frankfurt high duty cycle H⁻ ion source is described. The investigations have shown that the ion source is capable of producing a 120 mA H⁻ beam at an arc power of 47.5 kW (50 Hz, 1.2 ms, duty cycle = 6 %). This corresponds to an emission current density of 153 mA/cm^2. The required extraction voltage for a matched beam is 33 kV. A rough estimation indicates an excellent beam emittance of only $\varepsilon_{rms,\,norm} \sim 0.07\pi$ mm mrad.

REFERENCES

1. K. Volk et al., *A Compact High Brilliance Ion Source for RFQ Injection*, Proceedings of EPAC94, London, June 27 – July 1, 1994.
2. Leung, Ka-Ngo., *Rev. Sci. Instrum.* **69** (2), 998 (1998).

BEAM FORMATION AND ACCELERATION

Modeling of Negative Ion Transport in Hydrogen Ion Sources — Estimation of Extracted H⁻ Current

O. Fukumasa, T. Fujioka and T. Fukuchi

Department of Electrical and Electronic Engineering, Faculty of Engineering, Yamaguchi University, Tokiwadai 2-16-1, Ube 755-8611, Japan

Abstract. The H⁻ trajectory is calculated by numerically solving the 3D motion equation, while the collisional processes of destruction, of elastic and of charge exchange are handled by a Monte Carlo procedure. The energy of the H⁻ ions is reduced through both an elastic collision with H^+ and a charge exchange collision with H. Extraction probability of a negative ion produced at any location inside the source is discussed and estimated.

INTRODUCTION

Neutral beam injection based on the negative ion source is one of the most promising candidates for heating and current drive in future fusion reactors. To generate intense beams of negative ions and optimize the negative ion source, understanding all the processes which occur in the source is important [1]. Then, to study H⁻ production in a tandem two chamber system, we have used the simulation model (a zero-dimensional code) shown in Fig.1 [2-4]. H⁻ density is obtained as a function of plasma parameters by solving a set of particle balance equations for steady-state hydrogen discharge plasmas with or without cesium injection. For discussing pressure dependence of the extracted H⁻ current, we have also estimated the extracted H⁻ ions from the H⁻ ions in the second chamber, i.e. H⁻ (2), by taking into account stripping loss of H⁻ ions in the acceleration grid region [3, 4], but the negative ion transport in the source is not taken into account [5].

In this article, we will discuss the extraction of negative ions with the use of our model calculation [3, 4] and modeling of negative ion transport (a three-dimensional code)[5]. So, we investigate kinetic energy of extracted negative ions and also estimate extraction probability of negative ions produced at any point in the extraction region by numerically solving the 3D motion equation.

SIMULATION MODEL

In the present study, with using a coordinate system shown in Fig. 2, negative ion trajectory in the second chamber shown in Fig. 1 is calculated numerically. When a negative ion is produced, it moves inside the source until destruction or extraction. This trajectory is governed by electric and magnetic fields, respectively E and B, through the equation of motion. i.e. $M\, dv/dt = q(E + v \times B)$, where M, v and q are the ion mass, velocity and charge. Its trajectory also depends on all collision processes. The following destruction and charge exchange collisions are taken into account by the Monte Carlo method [5, 6] : (1)electronic detachment (ED) $H^- + e \rightarrow H + 2e$, (2)mutual neutralization (MN) $H^- + H^+ \rightarrow 2H$, (3) $H^- + H_2^+ \rightarrow H + H_2$, (4) $H^- + H_3^+ \rightarrow 2H + H_2$, (5)associative detachment (AD) $H^- + H \rightarrow H_2 + e$, (6) $H^- + H_2 \rightarrow H + H_2 + e$, (7) $H^- + Cs^+ \rightarrow H + Cs$, (8) $H^- + Cs \rightarrow H + Cs + e$, and (9) charge exchange (CX) $H^- + H \rightarrow H + H^-$ [7]. In addition to these collision processes, elastic collision with H^+ ion is also taken into account.

As is shown in Fig. 2, test volume-produced H^- ions are launched isotropically in all directions at any location with an initial energy of 0.5 eV except that z is set at four different points (i.e. 0.25, 0.75, 1.25 and 1.95 cm), and test surface-produced H^- ions are launched perpendicularly from the plasma grid with an initial energy of 0.5-2 eV, taking into account plasma sheath acceleration. The spatial profiles of the magnetic filter are given by the Gaussian profile $B_x(y, z) = B_0 \exp[-(z-z_0)^2/l_B^2]$, with $z_0 = 2.0$ cm, $l_B = 1$ cm and $B_0 = 120$ G.

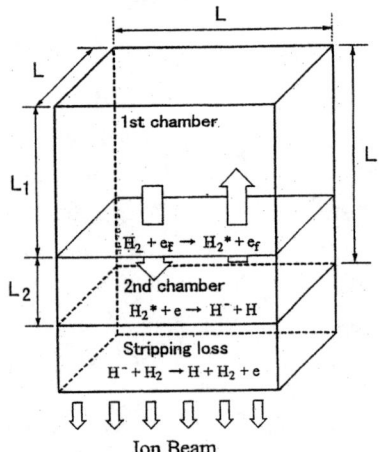

Fig. 1 Simulation model for the tandem two-chamber system.

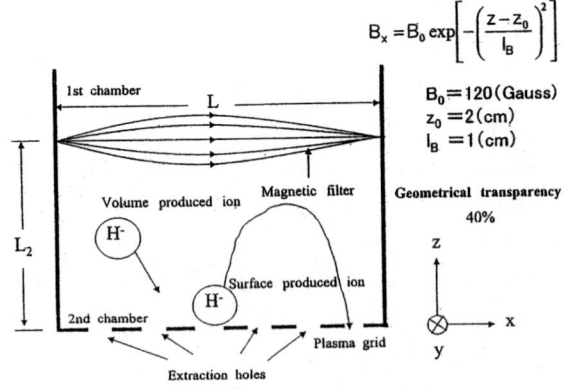

Fig. 2 Cross-sectional view of the model geometry for the second chamber of the tandem system shown in Fig. 1.

n_e	Electron density	1.00×10^{12} cm^{-3}
n_H	H atom density	5.22×10^{13} cm^{-3}
n_{H_2}	H$_2$ atom density	8.31×10^{13} cm^{-3}
n_{H^+}	H$^+$ ion density	3.73×10^{11} cm^{-3}
$n_{H_2^+}$	H$_2^+$ ion density	2.71×10^{11} cm^{-3}
$n_{H_3^+}$	H$_3^+$ ion density	1.52×10^{11} cm^{-3}
n_{Cs^+}	Cs$^+$ ion density	5.85×10^{11} cm^{-3}
n_{Cs}	Cs atom density	4.41×10^{12} cm^{-3}
T_e	Electron temperature	1.0 eV
T_H	H atom temperature	0.5 eV
T_{H^+}	H$^+$ ion temperature	0.5 eV

TABLE I Plasma parameters used in this simulation when gas pressure $p = 5$ mTorr.

The background plasma profiles are assumed to be uniform, and these values are obtained by the previous model calculation [3, 4] and are used to estimate mean free paths for collisions mentioned above. To determine the electron density dependence of H$^-$ production and particle densities, calculation is performed as a function of electron density $n_e(1)$ in the first chamber on the assumption that other plasma parameters are kept constant [2-4]. A certain numerical result is summarized in Table I. Plasma conditions for model calculation is as follows: the gas pressure $p = 5$ mTorr, the electron density ratio between two chambers $n_e(1)/n_e(2) = 0.2$, density of e_f in the first chamber $n_{fe}(1)/n_e(1) = 0.05$, electron temperature in the first and second chambers are, respectively, $\kappa Te(1) = 5$ eV, $\kappa Te(2) = 1$ eV, and magnetic filter position $L_1 : L_2 = 28 : 2$ cm (i.e. $z_0 = L_2 = 2$ cm).

NUMERICAL RESULTS AND DISCUSSION

The code calculates a negative ion trajectory until this ion is destroyed or extracted (reached to the plasma grid, PG). The motion equation is numerically solved by a Runge-Kutta-Gill method. A typical orbit of H⁻ ions in the second chamber of the negative ion source is shown in Fig. 3.

For a certain plasma conditions, a set of five calculations (one calculation for surface-produced H⁻ ions and four calculations for volume-produced H⁻ ions with different four z positions) is done. We typically used 10^3 test H⁻ ions for one calculation. A typical example is summarized in Table II. In the present case, 491 surface-produced H⁻ ions reached the PG and extraction probability is about 20 % (geometrical transparency of the PG is assumed to be 40 %). For volume-produced H⁻ ions, the probability to reach the PG decreases with upstream distance z from the PG. Then, mean value of the extraction probability is 15.7 %. This probability depends on gas pressure. As is shown in Fig. 4, extraction probability of volume-produced H⁻ ions decreases with gas pressure, but that of surface-produced H⁻ ions keeps a nearly constant value. For volume-produced H⁻ ions, the same calculation is carried out with initial energy of 0.2 eV. Extraction probability is kept nearly the same value with the case of 0.5 eV.

By the way, for simplicity, the modeling is made for a constant and equal plasma potential in the first and second chambers. With this choice of plasma potential, H⁻ ions are injected from the first chamber into the second chamber, but this effect is not considered in the present simulation and may modify a little the number of H⁻ ions reaching the PG. Namely, a little enhancement of extraction probability may be expected.

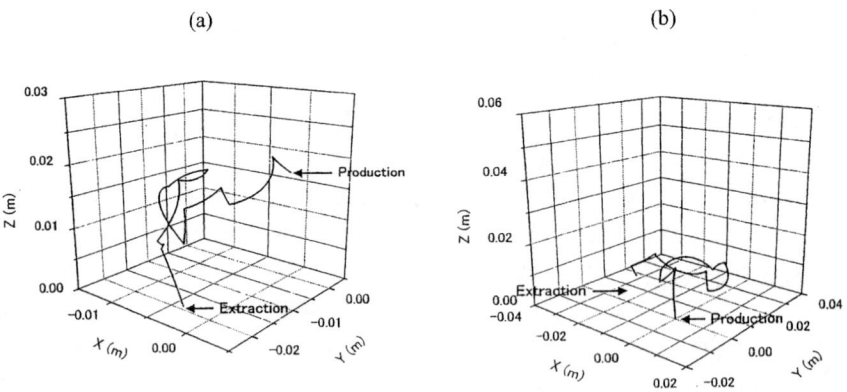

Fig. 3 Typical orbit of H⁻ ions in the second chamber : (a) a volume-produced H⁻ ion (initial energy : 0.5 eV, birth point (x, y, z) = (0, 0, 1.75 cm)), (b) a surface-produced H⁻ ion (initial energy : 1 eV, birth point (x, y, z) = (0, 0, 0)).

Kinds of H⁻ ion loss		Surface produced H⁻ ions	Volume produced H⁻ ions			
			Birth point of volume production [cm]			
			0.25	0.75	1.25	1.75
Wall loss		0	0	0	0	0
Flow out to the 1st chamber		233	141	231	355	581
Collisional destruction	e	11	8	13	7	12
	H^+	27	43	41	34	32
	H_2^+	22	21	35	48	28
	H_3^+	8	15	13	15	13
	H	52	51	65	58	36
	H_2	70	75	99	102	92
	Cs^+	77	36	31	40	28
	Cs	9	8	10	9	7
Total		276	257	307	313	248
Elastic collision	H^+	576	635	808	867	652
Charge exchange	H	1649	1286	1703	1735	1326
H⁻ ion reached the PG (extraction)		491	602	462	332	171
Extraction probability [%]		20.6	24.1	18.5	13.3	6.8

TABLE II Numerical results of H⁻ transport when $p = 5$ mTorr.

Energy of the extracted H⁻ ions is reduced compared with initial energy due to elastic collision with H^+ ions [8] and charge exchange collision with H [5]. For volume-produced H⁻ ions, mean value of an extracted H⁻ ion energy is reduced to 0.36 eV compared with initial energy 0.5 eV. On the other hand, when initial energy is set at 0.2 eV, mean value of the extracted H⁻ ion energy is increased up to about 0.3 eV. For surface-produced H⁻ ions, that is reduced to 0.37 eV compared with initial energy of 1 and 2 eV, and is reduced to 0.35 eV with initial energy of 0.5 eV, respectively. According to our simulation results, charge exchange collision plays an important role in energy relaxation of the extracted H⁻ ions, in particular the surface-produced H⁻ ions. This energy relaxation is the cause for good beam optics of negative ion source with cesium injection.

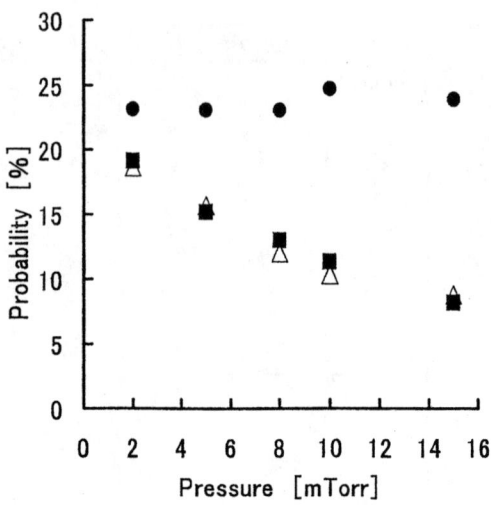

Fig. 4 Pressure dependence of the probability for H^- ions to reach the plasma grid : ● for surface-produced H^- ions, △ and ■ for volume-produced H^- ions with and no Cs.

CONCLUSIONS

The probability for H^- ions to reach the plasma grid is estimated. The extraction probability is relatively low. Within the present numerical conditions, the extraction probability for surface-produced H^- ions keeps nearly the constant value (i.e. 23-25 %), and that for volume-produced H^- ions decreases in its value from 18 % to 8 % with increasing gas pressure. The kinetic energy of the extracted H^- ions is reduced mainly through charge exchange collision with H. It is confirmed that energy of surface-produced H^- ions is nearly equal to that of volume-produced H^- ions in the vicinity of the plasma grid.

ACKNOWLEDGEMENTS

A part of this work was supported by the Grant-in-Aid for Scientific Research from the Ministry of Education, Culture, Sports, Science and Technology, Japan.

REFERENCES

1. Fukumasa, O., *J. Phys.* **D22**, 1668-1679 (1989).
2. Fukumasa, O., *J. Appl. Phys.* **71**, 3193-3196 (1992).
3. Fukumasa, O. and Monji, H., *Rev. Sci. Instrum.* **71**, 1234-1236 (2000).
4. Fukumasa, O., *IEEE T. Plasma Sci.* **28**, 1009-1015 (2000).
5. Riz, D. and Pamela, J., *Rev. Sci. Instrum.* **69**, 914-919 (1998).
6. Asano, S., Tsumori, K., Okuyama, T., Suzuki, Y., Osakabe, M., Oka, Y., Takeiri, Y. and Kaneko, O., *Rev. Sci. Instrum.* **70**, 2338-2344 (1999).
7. Hummer, D. G., Stebbings, R. F., Fite, W. L. and Branscomb, L. M., *Phys. Rev.* **119**, 668-670 (1960).
8. Makino, K., Sakurabayashi, T., Hatayama, A., Miyamoto, K. and Ogasawara, M., *Rev. Sci. Instrum.* **73**, 1051-1053 (2002).

An Inverted Plasma Sheath for the Simulation of the Extraction of Volume Produced H-

Reinard Becker

*Institut für Angewandte Physik der Johann Wolfgang Goethe-Universität
D-60054 Frankfurt/M, Fach 180, Germany*

Abstract. For the extraction of positive ions from plasmas well established computer programs are available, which are based on the Bohm sheath theory. In general the results of such simulations agree very well with experimental data. The situation is completely different, however, for the simulation of the extraction of volume produced H- ions. An open question so far has been, if there will be a saddle point of potentials in the extraction path or there will be an inverted sheath. No simple theory existed so far for the formation of an inverted sheath, connecting the quasi-neutral plasma in a self-consistent manner with the field provided by the positive extraction voltage. Based on the formulation of the space charge of the virtual cathode, caused by reflected protons in the extraction aperture, a linear model for an inverted plasma sheath will be presented, which allows to discuss the influence of most physical processes in the formation and extraction of H- and may become the basis of a correct simulation program.

INTRODUCTION

The extraction of negative ions, which are produced in the volume of plasma sources, has not been accessible to a physically sound numerical simulation, although simulations have been rather successful using computer programs for positive ion extraction just by interpreting voltages reversed and taking for the positive ion current the sum of negative ion current and space charge adjusted electron current with inversed sign [1].

A correct numerical simulation needs to overcome the space charge singularity caused by the positive ions, which are reflected by the applied field for the extraction of negative ions and may form a region of infinite space charge in free space. This is impossible in reality and blowing up any numerical treatment. Instead, a virtual cathode will develop and - as observed in experiments of decelerating electron beams [3] - the transported current is reduced in proportion to the 3/2 power of the voltage difference between birth potential of the positive ions and the local potential. Another important finding [4] is related to the emittance of extracted ions, suggesting that there is no potential minimum in the extraction channel, hence no saddle point with adverse optical influence, as assumed elsewhere [5]. It has been shown earlier[6] that a self consistent numerical simulation without a potential minimum does agree very well with these assumptions. In this paper a systematic evolution of a simplest formulation

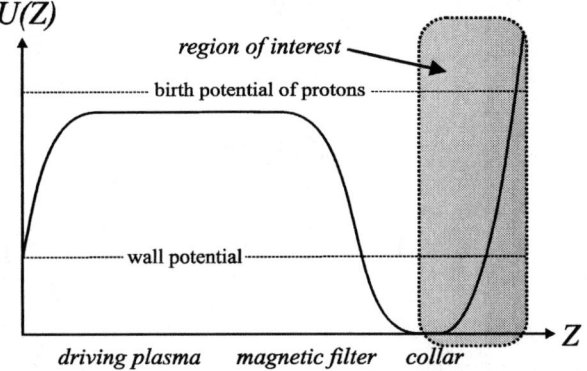

FIGURE 1. Schematic axial potential function in an ion source for volume production of H. according to Peters [2]: Left is the driving plasma, right the extraction region and in between the magnetic filter, which prevents the fast electrons to enter into the extraction region. In the collar region the plasma potential drops below the wall potential.

to a more and more complex one is performed, showing that only the presence of thermal positive ions – new born protons, ions from sputtered atoms of the plasma electrode or cesium ions - can provide sufficient positive space charge to establish a stable inverted sheath.

A VIRTUAL CATHODE EXPERIMENT RELATED TO H. - EXTRACTION

The reflection of protons in the extraction field for negative H.-ions is creating a virtual cathode situation [6]: In an experiment with an EBIS [3] it had been shown that the transported electron current through a decelerating potential drop is decreased in proportion to the 3/2 power of the voltage difference between the onset of decrease – called the virtual cathode potential U_V - and the birth potential U_p. This experimental result agrees well with the theoretical prediction for the potential of the virtual cathode, but the reduction of transported electron current is missing any simple understanding, as well as the 3/2 power law, which has been observed. Here, however, it can be used to describe the fading away of the current, current density, and space charge of protons, reflected in a virtual cathode:

$$j_p(U) = j_p(U_V) \times \left(\frac{U_p - U}{U_p - U_V} \right)^{3/2} \text{ for } U_V < U < U_p \tag{1}$$

by using the relation between current density and space density the current density as well as the space density can be described by corresponding values at $U = 0$, while the potential U_V drops out, if $U_V < U < U_p$:

$$j_p(U) = en_p(0)\sqrt{\frac{2e}{M_p}U_p} \times \left(1 - \frac{U}{U_p}\right)^{3/2}$$

and (2)

$$n_p(U) = n_p(0) \times \left(1 - \frac{U}{U_p}\right)$$

MODELLING OF AN INVERTED PLASMA SHEATH

With the prerequisite of "no saddle point", the mathematical physics for the simulations are becoming simple: electrons and negative hydrogen ions continuously are accelerated out of the neutralised plasma, protons being reflected by the extraction field. Their density increases by deceleration and is reduced by the fading away of the current density according to the virtual cathode formation for potentials above the virtual cathode potential U_v [6]. Although a more rigorous treatment should use thermal electrons and ions, we assume that their space charge is associated to their current density by uniform velocities inside the plasma, corresponding to stopping voltages U_- and U_e. This gives a reasonable physical description, as long as these temperature-equivalent stopping voltages are small as compared to the acceleration voltages applied and allows easy implementation in a program like IGUN [7]:

$$n_-(U) = \frac{n_-(0)}{\sqrt{1 + \frac{U}{U_-}}}, \quad n_e(U) = \frac{n_e(0)}{\sqrt{1 + \frac{U}{U_e}}}, \quad (3)$$

Depending on the relative position of U_V with respect to the plasma potential $U=0$ we have three different formulations for the ion density, while the H. and electron density are the same:

$$n_p(U) = \frac{n_p(0)}{\sqrt{1 - \frac{U}{U_p}}} \quad \text{for } 0 < U < U_V$$

$$n_p(U) = \frac{n_p(0)}{\sqrt{1 - \frac{U}{U_p}}} \times \left(\frac{U_p - U}{U_p - U_V}\right)^{3/2} \quad \text{for } 0 < U_V < U \quad (4)$$

$$n_p(U) = n_p(0)\left(1 - \frac{U}{U_p}\right) \quad \text{for } U_V < 0 < U$$

The first two formulations belong to the situation, where the potential for the development of the virtual cathode is above the plasma potential. In this case it can be shown, that integration of the Poisson equation is possible and that even a solution may be obtained with zero field strength at the transition from the plasma sheath to the extraction field. However, for small values of positive U the field strength becomes imaginary, which resembles the Bohm instability in the sheath formulation for positive ion extraction.

The third formulation considers the distributed reflection by a virtual cathode, as detected with our EBIS [3], taking into account the increase of the ion density by deceleration and its decrease by the virtual cathode action. Surprisingly enough, the potential for the onset of the virtual cathode drops out completely.

It is convenient to introduce the experimentally well known parameter of electron to H_ - current, which we will call Γ. From its definition (eq. 5) we obtain eq. 6:

$$\Gamma = \frac{j_e}{j_-} = \frac{en_e(0)\sqrt{\frac{2e}{m_e}U_e}}{en_-(0)\sqrt{\frac{2e}{M_p}U_-}} \tag{5}$$

$$n_e(0) = n_-(0)\,\Gamma\sqrt{\frac{m_e}{M_p}\frac{U_-}{U_e}} \tag{6}$$

with this and assuming quasineutrality at the plasma potential $U=0$

$$-\left(1+\Gamma\sqrt{\frac{m_e}{M_p}\frac{U_-}{U_e}}\right)n_-(0) + n_p(0) = 0 \tag{7}$$

all densities may be expressed by $n_{-,0}$, which is determined by the experimentally well known parameter of the H_ current density. Then it is the easy to integrate the Poisson equation and to show that the field strength for small values of U remains real and positive, necessary for establishing a stable inverted plasma sheath.

Similar to the positive ion extraction problem the potential of the plasma with respect to the wall is found by the balance of currents or current densities, respectively:

$$j_- = en_-(0)\sqrt{\frac{2e}{M_p}U_-}$$

$$j_e = en_e(0)\sqrt{\frac{2e}{m_e}U_e} \tag{8}$$

$$j_p(U_W) = en_p(0)\sqrt{\frac{2e}{M_p}U_p} \times \left(1 - \frac{U_W}{U_p}\right)^{3/2}$$

resulting in

$$\frac{U_W}{U_p} = 1 - \left(\frac{1+\Gamma}{1+\Gamma\sqrt{\frac{m_e}{M_p}\frac{U_-}{U_e}}}\right)^{2/3} \times \left(\frac{U_-}{U_p}\right)^{1/3} \tag{9}$$

In order to keep the wall potential between the plasma potential $U=0$ and the birth energy of protons, the term with the brackets may not exceed the value of 1. This is possible in the case of cesium seeding with $1<\Gamma<10$, but impossible without, where $10<<\Gamma$. In this case eq. 9 approximately becomes:

$$\frac{U_w}{U_p} = 1 - \left(\frac{M_p U_e}{m_e U_p}\right)^{1/3} \quad \text{for } 10<<\Gamma \tag{10}$$

It seems not realistic that the electron energy is less than 1/1836 of the proton energy, hence the wall potential will become negative with respect to the plasma potential. This, however, is violating the assumptions made for the existence of an inverted sheath, because by a positive plasma potential with respect to the wall potential, a saddle point will develop in the extraction aperture.

POSITIVE IONS IN THE NEGATIVE PLASMA CHANNEL

Looking at a negative plasma channel in front of the extraction aperture it appears most adequate to also take into account a thermal distribution of positive ions, which are born from gas by ionizing collisions of the fast protons. These ions can consist of background gas, sputtered atoms of the extraction electrode or cesium, or of a combination of all. Since they are positive, they are trapped in 3 dimensions by the increase of the potential towards the main plasma chamber, towards the extraction aperture and – in case of a collar – even radially. The space charge contribution of theses ions will be given by a Boltzmann distribution, relating their axial density to the density of H. - ions by δ, which is unknown for the moment, and by U_+, which is a stopping voltage equivalent to a temperature:

$$n_+(U) = \delta n_-(0) \exp\left\{-\frac{U}{U_+}\right\} \tag{11}$$

The condition of quasineutrality at $U=0$ now modifies eq. 7 to

$$n_p(0) = n_-(0)\left(1 + \Gamma\sqrt{\frac{m_e U_-}{M_p U_e}} - \delta\right). \tag{12}$$

With this we can write for the one-dimensional Poisson equation:

$$\frac{d^2 U}{dz^2} = \frac{en_-(0)}{\varepsilon_0}\left\{\frac{1}{\sqrt{1+\frac{U}{U_-}}} + \frac{\Gamma\sqrt{\frac{m_e U_-}{M_p U_e}}}{\sqrt{1+\frac{U}{U_-}}} - \left(1+\Gamma\sqrt{\frac{m_e U_-}{M_p U_e}} - \delta\right)\left(1 - \frac{U}{U_p}\right) - \delta\exp\left\{-\frac{U}{U_+}\right\}\right\} \tag{13}$$

which can be integrated and added with a constant to assure that the field strength becomes zero at $U=0$:

$$\left(\frac{dU}{dz}\right)^2 = \frac{en_-(0)}{\varepsilon_0} \left\{ \begin{array}{l} 2U_-\left[\sqrt{1-\frac{U}{U_-}}-1\right]+2U_e\left[\sqrt{1-\frac{U}{U_e}}-1\right]\Gamma\sqrt{\frac{m_e U_-}{M_p U_e}} \\ +\left[1+\Gamma\sqrt{\frac{m_e U_-}{M_p U_e}}-\delta\right]\frac{U^2}{2U_p}+\delta\left[\exp\left\{-\frac{U}{U_+}\right\}\right] \end{array} \right\} \quad (14)$$

For small values of U the expression on the right side must remain positive, which needs development of the square-root and the exponential function to first order, resulting in

$$0 < 1+\Gamma\sqrt{\frac{m_e U_-}{M_p U_e}} - \delta \quad (15)$$

limiting the amount of trapped positive charges to assure the equivalent Bohm criterion for a stable sheath. In addition to wall currents of H., electrons and protons, as described in eq. (8) now also the trapped positive charges contribute to the balance of currents, establishing the plasma potential:

$$j_+ = en_-(0)\sqrt{\frac{2M_+}{\pi kT_+}} \times \int_{\sqrt{2eU_W/M_+}}^{\infty} v\exp\{-M_+v^2/2kT_+\}dv \quad (16)$$

Assuming $U=0$ at the beginning of the sheath (see Fig. 1), the wall potential is found by solving the following transcendent equation for the balance of all wall currents in dependence of the unknown parameter δ:

$$1+\Gamma = \left(1+\Gamma\sqrt{\frac{m_e U_-}{M_p U_e}} - \delta\right)\sqrt{\frac{U_p}{U_-}}\left(1-\frac{U_W}{U_p}\right)^{3/2} + \delta\sqrt{\frac{M_p U_+}{M_+ U_-}}\exp\left\{-\frac{U_W}{U_+}\right\} \quad (17)$$

NUMERICAL EVALUATION AND DISCUSSION

A simple Newton procedure has been written in order to find numerical solutions to eq. 17. For appropriate values of the parameters the following assumptions have been made: $U_-=U_+=U_e=1$ and $U_P=15$, assuming a higher electron temperature in the driving plasma. Values of $\Gamma=1, 2, 3$ were used in combination with $M_P=133$, reflecting the case of cesium seeding, while high values of $\Gamma=50, 100, 200$ were used together with $M_P=1$. Starting at the maximum value of δ as given by eq. 15, those values of U_W were searched for, which satisfy eq. 17. This results in the curves shown in Fig. 2. Falling curves belong to the case of cesium seeding, the raising ones to the case without. A positive value of U_W and hence the possibility for the existence of an inverted sheath is only found in the case of cesium seeding at rather low values of Γ. For higher values of Γ – in the case of no cesium added – no inverted sheath can exist. A bias applied to the extraction electrode or to the collar with respect to the body of the source at first glance will not change the presented picture, however, since the

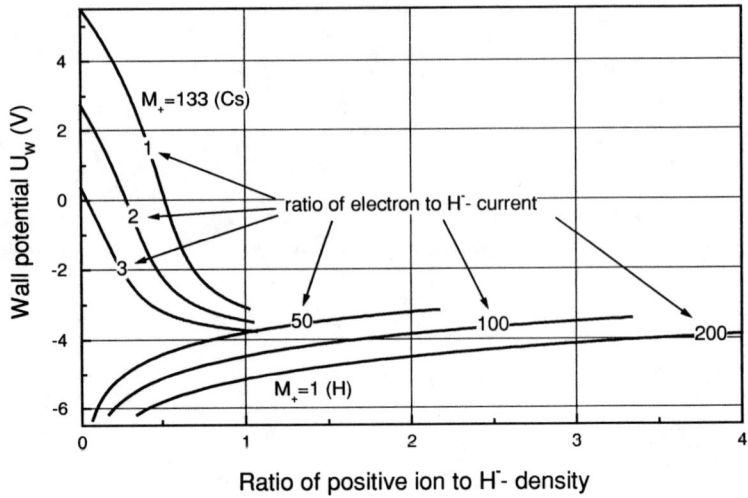

FIGURE 2. Wall potential U_W according to eq. 17 for $U_-=U_+=U_e=1$ and $U_P=15$ in dependence of the ratio of positive ion to H_- density (δ). Parameter is the electron to H_- current (Γ) and according to values of $\Gamma=1,2,3$ (cesium) $M_P=133$ and for $\Gamma=50,100,200$ (hydrogen ions) $M_P=1$.

electron current will be reduced directly by a negative bias, other parameter ranges may become possible.

CONCLUSIONS

A mathematical model has been developed for an inverted sheath in the case of volume produced negative hydrogen ions. It has been shown that the virtual cathode formed by the reflection of protons in the extraction field for negative ion is well suited to build up an inverted plasma sheath, using the experimental results from transmitted electron currents in an electron beam ion source (EBIS). A potential increase from the plasma to the wall electrode can be only achieved for a low ratio of electron to H_current. This sheath formulation therefore is restricted to the case of cesium seeding. The results presented here will be used to develop a realistic negative hydrogen extraction simulation program, based on IGUN [7]. Experiments then will be needed to show the necessity for further refinements, like thermal initial distributions for electrons and negative hydrogen ions.

REFERENCES

1 Leitner, M. et al., *Nucl. Instrum. and Meth.*, **A 427,** 242 (1999)
2 Peters, J., *Rev. Sci. Instrum.* **73,2** ,900 (2002)
3 Kleinod, M. et al., *Proc. SCHEF'99*, Dubna 1999, p. 129, edts : E.D. Donets, E.E. Donets, I. Meshkov
4 Leung, K. et al., *Rev. Sci. Instrum.* **54,** 56 (1983)
5 Whealton, J. et al., *Rev. Sci. Instrum.*, **69,** 1103 (1998)
6 Becker, R., *Rev. Sci. Instrum.*, **69,** 1107 (1998)
7 Becker, R. and Herrmannsfeldt, W.B., *Rev. Sci.Instrum.* **63** 2756 (1993)

Studies on the Extraction Region of the Type VI RF Driven H⁻ Ion Source

P. McNeely[1], M. Bandyopadhyay, P. Franzen, B. Heinemann, C. Hu,
W. Kraus, R. Riedl, E. Speth, and R. Wilhelm

Max-Planck-Institut für Plasmaphysik
EURATOM Association
Boltzmannstrasse 2
Garching, D-85748
Germany

Abstract. IPP Garching has spent several years developing a RF driven H⁻ ion source intended to be an alternative to the current ITER (International Thermonuclear Experimental Reactor) reference design ion source. A RF driven source offers a number of advantages to ITER in terms of reduced costs and maintenance requirements. Although the RF driven ion source has shown itself to be competitive with a standard arc filament ion source for positive ions many questions still remain on the physics behind the production of the H⁻ ion beam extracted from the source. With the improvements that have been implemented to the BATMAN (Bavarian Test Machine for Negative Ions) facility over the last two years it is now possible to study both the extracted ion beam and the plasma in the vicinity of the extraction grid in greater detail.

This paper will show the effect of changing the extraction and acceleration voltage on both the current and shape of the beam as measured on the calorimeter some 1.5 m downstream from the source. The extraction voltage required to operate in the plasma limit is 3 kV. The perveance optimum for the extraction system was determined to be 2.2×10^{-6} A/V$^{3/2}$ and occurs at 2.7 kV extraction voltage. The horizontal and vertical beam half widths vary as a function of the extracted ion current and the horizontal half width is generally smaller than the vertical. The effect of reducing the co-extracted electron current via plasma grid biasing on the H⁻ current extractable and the beam profile from the source is shown. It is possible in the case of a silver contaminated plasma to reduce the co-extracted electron current to 20% of the initial value by applying a bias of 12 V. In the case where argon is present in the plasma, biasing is observed to have minimal effect on the beam half width but in a pure hydrogen plasma the beam half width increases as the bias voltage increases. New Langmuir probe studies that have been carried out parallel to the plasma grid (in the vicinity of the peak of the external magnetic filter field) and

---

[1] Email: p.mcneely@ipp.mpg.de
Fax: 49 89 3299 2558
Phone: 49 89 3299 1339

changes to source parameters as a function of power, and argon addition are reported. The behaviour of the electron density is different when the plasma is argon seeded showing a strong increase with RF power. The plasma potential is decreased by 2 V when argon is added to the plasma. The effect of the presence of unwanted silver sputtered from the Faraday screen by Ar^+ ions on both the source performance and the plasma parameters is also presented. The silver dramatically downgraded source performance in terms of current density and produced an early saturation of current with applied RF power. Recently, collaboration was begun with the Technical University of Augsburg to perform spectroscopic measurements on the Type VI ion source. The final results of this analysis are not yet ready but some interesting initial observations on the gas temperature, disassociation degree and impurity ions will be presented.

I. Introduction

The Neutral Beam Group at IPP has for the past five years been developing a RF based negative Hydrogen ion source as an alternative to a filamented arc ion source (the current ITER reference source). This work has been collaboration with CEA Caderache in the past, and is currently being formalised under an EFDA contract. The Type VI ion source reported on in this paper has been previously described in detail[1] and shown in Figure 1 is a diagram of the source.

Of particular importance in this paper is the extraction system. It was provided by CEA Cadarache and is copy of a JAERI design. There are 49 apertures arranged in a 7 x 7 rectangle. The apertures are 14 mm in diameter. Five of the apertures are masked off leaving a total extraction area of 67.7 cm^2. The aperture spacing centre to centre is 19 mm in the horizontal direction and 21 mm in the vertical direction. The plasma grid is 2 mm thick molybdenum. The extraction gap is 4.5 mm. The water-cooled copper extraction grid is 11 mm thick and the apertures are 12 mm in diameter. The acceleration gap is 8.2 mm. The copper grounded grid has a thickness of 3 mm with 14 mm diameter apertures. In the past attention was focused on the source's performance primarily in terms of H$^-$ current density. This was due to the goals of the research effort but was also in part due to a lack of diagnostics. In the year 2001 improvements to the data acquisition (DAQ) with the calorimeter significantly enhanced the ability to analyse the properties of the extracted ion beam. A CAMAC based DAQ was installed in the summer of 2001. This system and its associated data evaluation software are based on the ASDEX-Upgrade Neutral Beam control system [2].

The advantages of this system are:
1. the complete calorimeter thermocouple signals (TC) are stored for each shot allowing the profile of the beam to be analysed.
2. all relevant source and test stand operational parameters (pressure, RF power, gas setting, etc) are stored.
3. automatic data logging allowing later analysis.

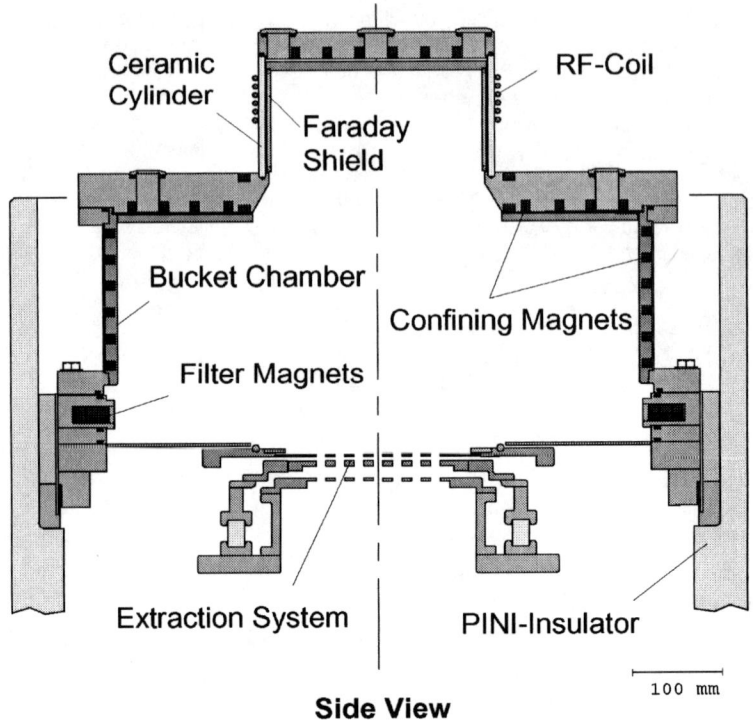

Figure 1. Above is drawing of the Type VI ion source. Shown are all major components of the ion source and extraction system.

The calorimeter itself was provided by CEA Cadarache and was 310 mm wide by 620 mm high. It consists of a 6x12 array of thermally isolated copper plates instrumented with thermocouples. Each plate is 51 x 51 mm^2 and 5 mm thick.

Evaluation of the beam current can now be based on either the sum of the calorimeter TC signals (as has been done in the past) or by fitting the thermocouple signals to a double Gaussian function. For all current densities reported in this paper the method of fitting the thermocouple signals was adopted as this accounts for the part of the beam not intercepted by the calorimeter.

Estimates have been performed [3] to determine the stripping losses in the accelerator system and in the main vacuum tank. For standard conditions (0.5 Pa Hydrogen and 25% Argon) the stripping losses in the accelerator are 25±5% for a 3 second pulse. The pulse length is important in BATMAN as a combination of Ti-Sorption and turbo-molecular pumps provide pumping in the main tank. Only the turbo-molecular pumps have any capacity to pump argon and their pumping speed (~ 5000 l/s in total) is insufficient to deal with the gas load from the source so over time the pressure in the tank rises to relatively high values (1x10^{-4} mbar).

A new magnetic filter flange was installed in November 2001. This filter flange has 10 KF-25 ports. They allow Langmuir Probe or other diagnostic measure-

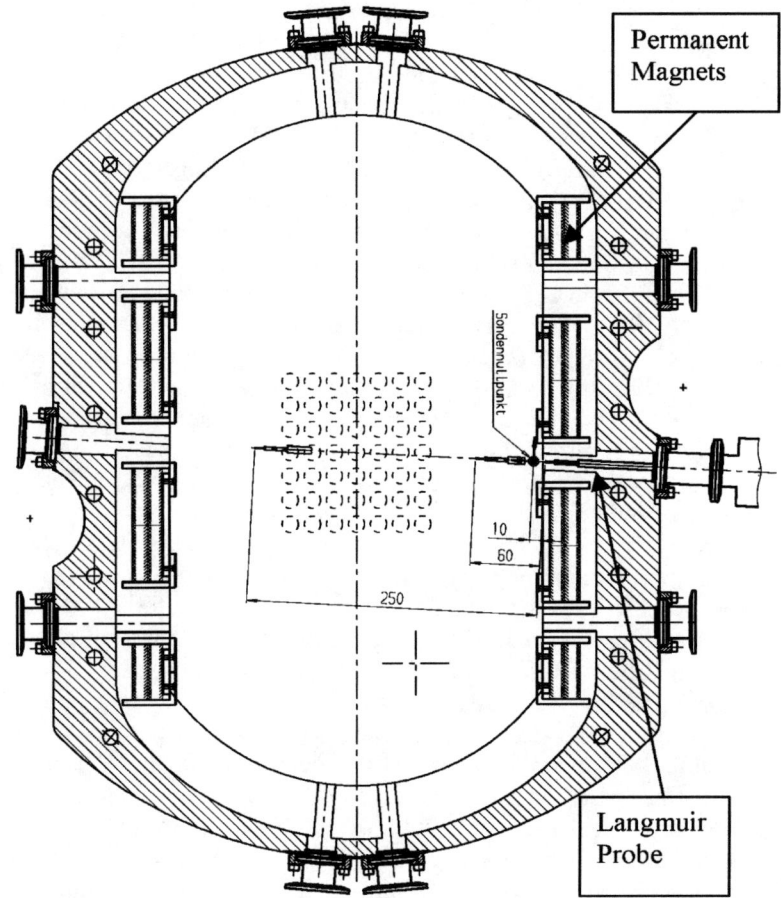

Figure 2. Shown above is drawing of the new diagnostic magnetic filter flange. The positions of the magnets in the flange itself are shown. The position of the extraction apertures are shown as broken circles. The horizontal Langmuir probes scan path is shown as is the position (indicated by a large cross) where the z-profile is collected by the Langmuir probe mounted on the back plate of the source. All dimensions shown are in mm.

ments in the vicinity of the plasma grid (see Fig. 2). A collaboration has begun with the TU Augsburg to make spectroscopic measurements of the ion source. In late 2001 the initial measurements have been carried out and data evaluation is complete, with a second experimental campaign in June 2002. The initial evaluation has clearly shown that sputtering is caused by argon ions as only when argon is added are AgI and CuI-lines present in the plasma. Also measured was the neutral gas temperature: 1200±300 K. The results of the second campaign have given more quantitative numbers on the disassociation degree (≈10%) and vibrational state distribution of the hydrogen molecules [4].

Accurate interpretation of the spectroscopic results requires that the electron density along the line of sight of the spectrometer be known. Thus Langmuir probe measurements in the extraction region have been performed with a new scanning probe and the APS3 hardware [5] from the older probe. The presence of the Faraday shield on the Type VI source has simplified probe measurement considerably. It is now possible to extend moderate power (<120 kW) pulses to 6 seconds (or longer) thus allowing for multiple collection of probe I-V characteristics improving statistical accuracy. It is still impossible to conduct Langmuir probe measurements in the driver region as the APS3 system has a current limit of 100 mA. Thus z-profile scans are still limited to a RF power of 60 kW, as previously [6]. As even at 60 kW the z-profile measurements with the old Langmuir probe (see Fig 10) are subject to analysis failure due excessive current, the length of the probe tip (made from 50 μm tungsten wire) was reduced from 5 mm to 2.6 mm. However, horizontal scans through the extraction region have been performed successfully up to 120 kW. This is a valuable enhancement of the Langmuir probe diagnostic capabilities.

The possibility of sputtering material from the Faraday shield when argon is added to the plasma has long been a concern. The amount of copper sputtered under normal operations, however; has never been excessive although deposited copper is always found on the plasma grid during routine maintenance. It does raise concerns when investigating different plasma grid materials as several showed substantial degradation in performance after the addition of argon when again operated purely with hydrogen[10]. The final section of the paper will present the results observed due to unintentional Silver contamination of the ion source.

The contamination came about due to a decision made to coat the Faraday shield with silver based on the results of our initial investigation into the effects of changing the plasma grid material on H⁻ current density [7]. The results showed copper contamination on the plasma grid (from operation with argon) and it was decided that to prevent this in the future the Faraday shield was to be coated with the same material as the plasma grid. As silver has a low workfunction when coated with caesium (<2 eV) it had been chosen as the first grid material to be tested in the new grid material campaign. During the initial phase of the campaign (operation without Cs) it became obvious that the source performance had become degraded. It was during this time as well that the Augsburg group made their initial spectroscopic measurements. The initial results of those measurements have conclusively shown that copper signals are only observed when argon is present.

By early 2002 the degradation in performance, with the silver grid and Faraday shield, was so severe that the source was opened to investigate. It was found that the entire inner surface of the source was covered in silver and that the silver coating of the Faraday shield was in places so thin that the underlying copper was visible. The source was then cleaned and silver contamination was removed from the surfaces and from the Faraday shield. After re-assembly the source performance improved but still did not achieve previous values. The source was again dismounted from the test stand

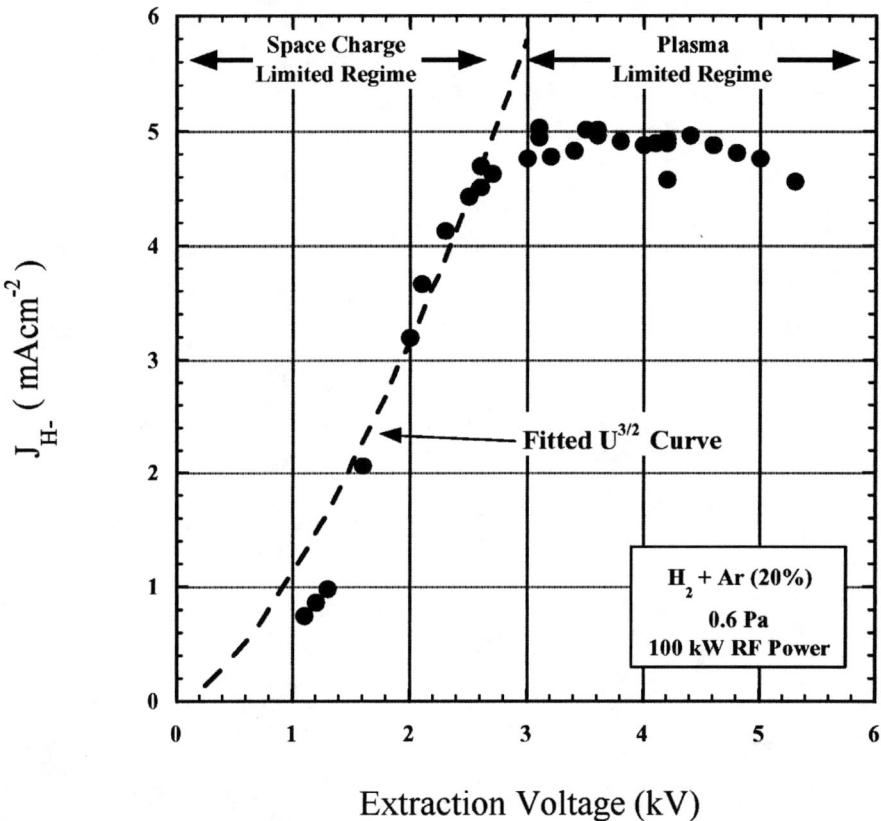

Figure 3. Shown is the H⁻ current density as a function of the extraction voltage for a silver contaminated plasma, with argon admixture and a pressure of 0.6 Pa at 100 kW of applied RF power. Shown on the graph is a $U^{3/2}$ curve fitted to the data.

and again substantial silver coating of the inner surface was found. It is not clear from where the silver came, although, the most likely place was the surfaces of the slots in the Faraday shield as these surfaces could not be cleaned mechanically.

The decision was taken to completely disassemble the source and rigorously clean every part plus the Faraday screen was to be recoated in a fresh layer of copper. The source cleaning, unfortunately, proved to be a difficult and time-consuming procedure due in large part to the similar chemical properties of silver and copper. In mid-May of 2002 the source was again mounted on the test bed in its "standard configuration" of Mo plasma grid and copper Faraday shield but with the new diagnostic flange and operation has again commenced.

II. Extraction Studies

The extraction system used with the Type VI source was provided to IPP by CAE Cadarache and is based on a JAERI design. It is identical to the system used initially on the KAMABOKO ion source tested on the MANTIS testbed [8]. SLAC

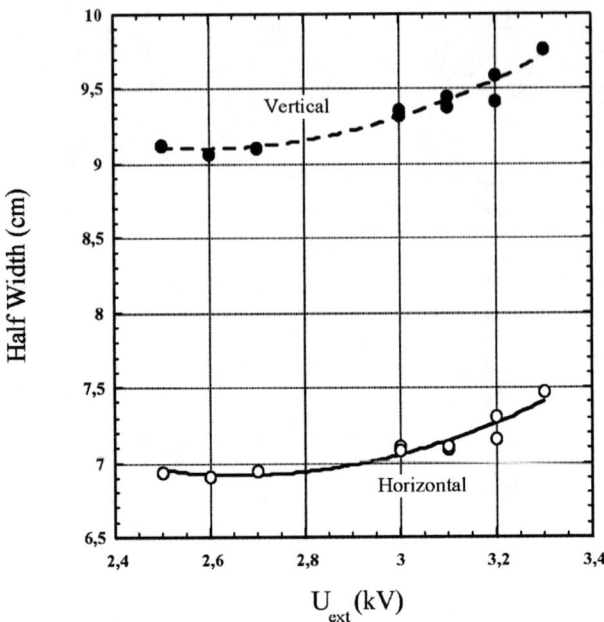

Figure 4. Shown is the beam half width (σ from the gaussian fit procedure) as a function of the extraction voltage for a constant total applied voltage to the voltage divider (17.5 kV) and current density for a silver contaminated plasma. The perveance optimum can be seen to be at 2.7 kV for the source from this graph.

code calculations [9] showed that the quality of the beam was dependant on both the choice of extraction/acceleration voltage and the current density of the extracted ion beam. In the past the data from the calorimeter could not be used to analyse the beam profile due to the lack of a storage system for the data from all 72 thermocouples. Using the new CAMAC based DAQ and associated analysis software it is now a simple matter to do so.

As the high voltage used in the extraction system comes from a single high voltage regulation tetrode a voltage divider is used to set the extraction and acceleration voltages. The voltage divider has a total resistance of 1 kΩ and the first resistance can be set to a value between 600 and 900 Ω (in steps of 50 Ω). A standard scan thus consists of shots at a specific voltage divider setting for different values of total applied high voltage.

In Fig. 3 can be seen the H⁻ current density as a function of the extraction voltage (U_{ext}). It is clear from this graph that for values of U_{ext} above 3 kV the source is operating in the plasma limited regime and the maximum possible H⁻ current is being extracted. The beam in the space charge limited region is reasonably close to the perveance optimum but no attempt to optimise divergence was made for this data. The deviation from the fitted curve at very low U_{ext} is likely due to bad beam optics.

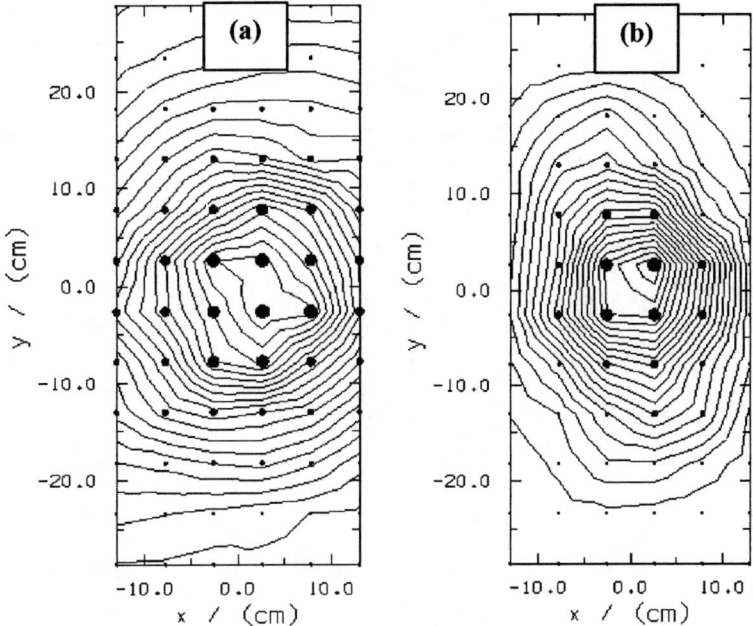

Figure 5. Shown above are two fitted profiles of the calorimeter for two shots. In (a) is shown a pure hydrogen discharge at 0.5 Pa and 44 kW RF power for a J_{H^-} of 1.9 mAcm^{-2} while (b) has H_2 plus 20% argon admixed at 0.8 Pa and 90 kW RF power for a J_{H^-} of 5.0 mAcm^{-2}. In both cases the source was contaminated by silver.

Simulation shows a substantial amount of the beam intercepts the grids when the beam current density is in this range [9] but it could also be due to a systematic error in the evaluation of the current density for these low values. As no bias voltage was used to suppress the co-extracted electrons for this scan it is possible that the H$^-$ current could be enhanced if the co- extracted electron current is reduced due decreased numbers of electrons available to fill the space charge limited current [9].

Figure 4 shows the beam widths for a constant total applied voltage (17.5 kV) and constant current density (plasma limited regime) to the voltage divider but differing extraction voltages for the same source conditions as in Fig. 3. It can be seen that the perveance optimum is found at 2.7 kV extraction voltage and corresponds to a perveance of 2.2×10^{-6} A/V$^{3/2}$. This is approximately 85% of the theoretical perveance for this extraction geometry (2.6×10^{-6} A/V$^{3/2}$). For the RF driven H$^+$ Sources (Type II) used on ASDEX-Upgrade the optimum perveance is 3.2×10^{-6} A/V$^{3/2}$. As the difference between the optimum perveance of the sources can be accounted for by the geometric differences between the extraction systems themselves this confirms that the beam dynamics of positive and negative sources is the same.

By fitting the measured profile to a double Gaussian function the dependence of the beam width in either the horizontal or vertical directions with the

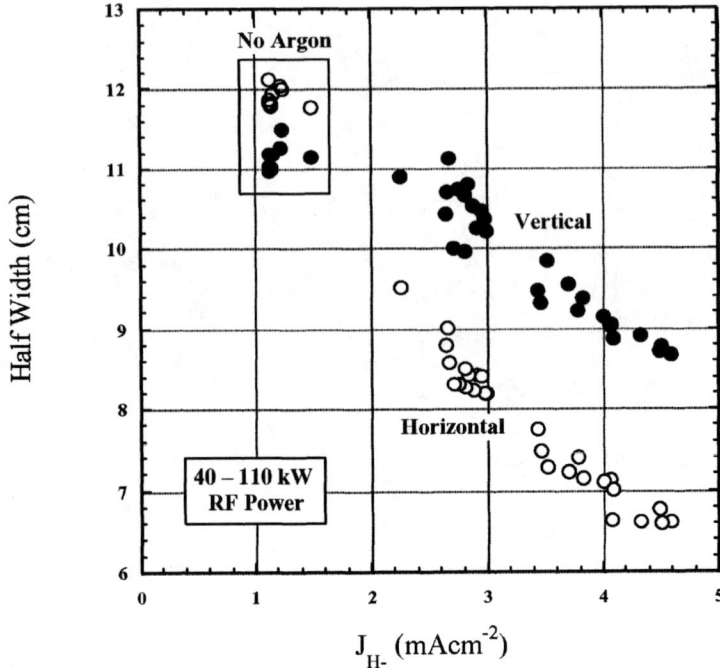

Figure 6. Shown is the horizontal (open circles) and vertical (solid circles) half widths of a Gaussian fit to the thermocouple signals of the calorimeter as a function of H⁻ current density. The source was contaminated by silver. For all shots the voltage applied to the divider was constant (17.5 kV), but the extraction voltage varied (2.5±0.5 kV). Argon (20%) was present except where indicated and the source pressure was in the range 0.5 to 0.8 Pa.

extraction/acceleration voltage or beam current density can be determined. In the past it had been qualitatively observed that as the beam current density increased the horizontal width of the beam was reduced. Quantitatively now this can be seen from the data (see Fig. 5). This figure shows the two profiles of the beam for different beam current densities and it is clear that the outer beam envelope is reduced as the beam current density increases and that the central region of the beam is more compact. What is also interesting is that the outer beam envelope is "rotated" 90° with respect to the horizontal in the case the higher current density shot.

The beam half width as a function of current density is shown in Fig. 6 and these results are in qualitative agreement with the results of the SLAC code calculation that predicts a decreasing beam divergence for increasing beam current density in the range we have measured. The validity of fitting with a Gaussian shape for the beam has been confirmed by using the DENSB code only taking into account geometric factors. For a divergence of 3 degrees and the small number of beamlets of the source the code predicts an essentially Gaussian beam at the calorimeter position as compared to a rectangular beam for 0.5 degrees divergence. An interesting effect is observed in Fig. 6. As the extracted current density increases the ratio of horizontal to vertical half width changes from >1 to <1. The reason for the substantial difference in the horizontal and vertical widths is not clear at this time. There exists no mechanical misalignment of the grids themselves. From the DENSB calculations the difference in horizontal and vertical spacing of the apertures is, alone, not sufficient to account for the observation as the code predicts a difference in the half widths only on the order of 3%. The only other known non-uniformity is the presence of the electron suppression magnetic field of the extraction grid superimposed on the main filter field. Calculations indicate that the electron suppression magnetic field is substantial (200 G) as compared to the magnetic filter field (30 G) at the plasma grid and that suppression magnetic field has a direction that alternates depending on the aperture considered. The presence of this combined field will clearly affect, strongly, the initial ion trajectories and the formation of the plasma miniscus but to determine the details quantitatively a full 3D analysis would be required. As the future of this grid is uncertain, a changeover to a new grid with a low penetration magnetic field is already planned, the effort can not be justified at this time.

For all shots used in Fig. 6 the voltage applied to the divider was constant (17.5 kV), however; there is a variation of the extraction voltage (2.5 ± 0.5 kV). This variation is due to higher H- ion currents having higher co-extracted electron currents and these currents change the current balance in the voltage divider. Due to this selection of extraction parameters the source was not operating in the plasma limit and so was not producing the maximum possible beam current (see Fig. 4) but was near to the optimum perveance for the extraction grids.

Although the new DAQ considerably improved the quality of the results obtained from the BATMAN calorimeter, the calorimeter itself still suffered from a number of limitations. As it lacked sufficient horizontal width to intercept the whole beam, fitting and extrapolation were required which is less than totally satisfactory especially given how critical an absolute measurement of the current density is. The calorimeter is only radiation cooled, which limits it to moderate pulse lengths only, and requires that the change in heat capacity with the temperature of the copper be accounted for.

It has been decided to replace the BATMAN calorimeter with one designed and built at IPP. The new calorimeter will be made of a 600 mm x 600 mm x 2 mm thick copper plate into which are inserted 144 carbon plugs. The central horizontal and vertical rows of the plugs will be instrumented with thermocouples that are read out by the DAQ. Onto the edge of the calorimeter is brazed a water-cooling pipe. The total beam power will be determined by water calorimetry. In addition the rear of the calorimeter can be viewed with an IR camera. The temperatures of the plugs should provide sufficient information to construct a beam profile. The calorimeter is currently under construction and will be installed mid-July 2002.

III. Effect of Biasing the Plasma Grid

Biasing experiments have been carried out and the results are shown in Fig. 7. For the first time with a RF source an enhancement of the H⁻ current density at low bias voltage as reported by Whealton and co-workers for an arc filament source [10] was observed. Moreover, most likely due to the presence of silver in the plasma it was possible to suppress the electron current with a lower bias voltage than was required in the past [1]. Previously, suppressing the co-extracted electron current by 50% required 22 V [1] of bias voltage and now it is possible to do so with only 8 V. Now either with pure hydrogen or hydrogen and argon a bias voltage of 12 V reduces the electron current on the extraction grid to 20% of its initial value. Comparing the results for horizontal width with and without argon the variation of the horizontal width appears

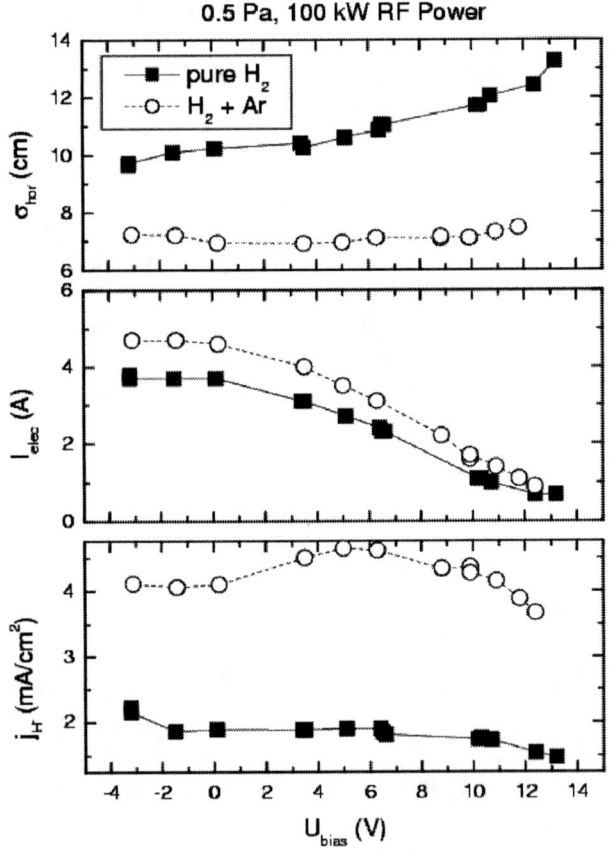

Figure 7. The results of biasing the plasma grid for a silver contaminated pure hydrogen and hydrogen plus 20% argon discharge at 0.5 Pa and 100 kW applied RF power.

to be correlated with the results shown in Fig. 6 for the pure hydrogen case: as the beam current decreases due to biasing the beam width increases. In the case of argon addition the effect is less pronounced.

IV. Langmuir Probe Measurements in the Extraction Region

A great deal of information could be gained from the Langmuir probe used in the past. However, it could only be operated under source parameters which were far from standard operating parameters and the probe sampled an area of plasma which was itself of little significant interest being some distance from the extraction region (see Fig 1 and 8). A second probe (functionally identical to the first) was purchased for use in the newly constructed "Diagnostic Magnetic Filter Flange" but installation was delayed due to the source being cleaned from silver contamination. The probe was installed in mid-May 2002, and the first tests (reported here) were highly successful. The location of the probe and its scan path can be found in Fig. 1 and Fig. 8. There is no obvious degradation of probe performance due to measuring within the magnetic filter field and due to the density of the plasma in this region measurements are possible up to 120 kW and probably somewhat higher in power, although this has not yet been tested. This is a critical finding as now the probe can be used to probe the plasma under standard source conditions.

The I-V curves collected from the probe are near to text book perfect (see Fig 9) and it is possible to use the collected EEDF (Electron Energy Distribution Function) to both determine the electron density (n_e) and the average electron energy ($<E>$). The plasma potential (U_{pl}) is determined by the zero-crossing point of the second derivative of the probe characteristic (as shown in Fig 9). In general two scan runs are performed at each source parameter setting and the results are averaged in the graphs presented in the paper. During each scan the probe moves to a pre-set starting point and then proceeds in 9 mm steps to the end point. At each step the probe stops, pauses 100 μsec, and then collects 3 complete IV curves that are automatically averaged. This is a departure from the procedure used in the past where the probe collected a single IV curve while moving. Although these changes to the probe collection procedure reduce possible errors it would still be prudent to assume that: all electron densities have a 10% error, all average electron energies have a 0.25 eV error, and there is a similar value for the error in the plasma potential.

Initial experiments with the old probe, mounted to perform a z-profile scan, showed that there is a difference between the electron density obtained if the probe scans during movement or after stopping (see Fig. 10). It is difficult to assess which method is more correct without another diagnostic capable of giving the electron density but given the complicated magnetic field the probe is moving through intuitively the scan performed after movement is complete is less suspect. It has been decided to implement a "collect I-V trace only after probe stops" procedure with all future probe measurements. Furthermore the measurement showed that the plasma has returned to the same state that it had before the silver contamination, and there is evidence the silver contamination resulted in changes to the plasma conditions.

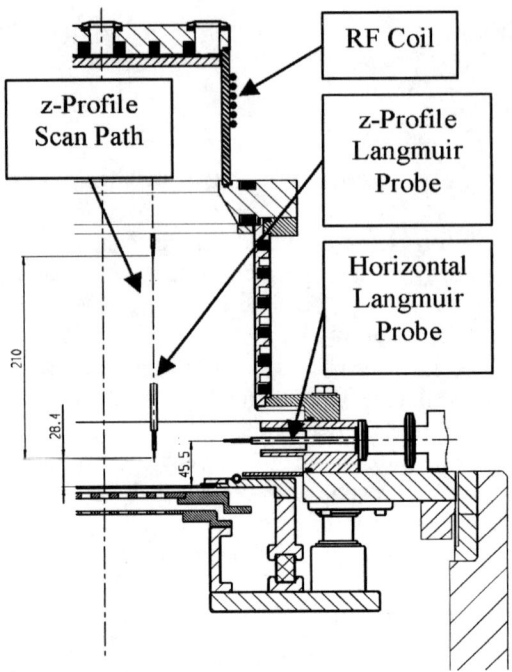

Figure 8. Shown is a drawing of the Type VI ion source indicating the position of the two Langmuir probes plus the scan path of the z-profile probe. All dimensions shown are in mm.

An important question about the plasma over the extraction region is the question of its uniformity. As can be seen in Fig. 11, which shows the electron density as a function of position (the centre of the plasma grid is approximately probe position 15 cm) the plasma is not uniform which is not surprising given the shape of the magnetic filter field (see Fig. 12). Over the region of the extraction apertures the plasma is reasonably uniform (the edges are within 10% of the central value) but shows a rapid and significant fall outside this region. Comparing Fig 12 to Fig 11 shows that this is due to an equally rapid rise in the strength of the magnetic filter field. As well, in central region between the 10 and 18 cm scan positions, all scans show an increase in the plasma density. As this effect exists in all scans it cannot be a simple measurement artefact, but it may be caused by the probe disturbing the plasma. However, in that case a decrease in plasma density would be the expected result. The effect is well correlated with the flattest part of the magnetic field profile (see Fig 12) and so could also be due to non-uniformity's that exist in the actual filter field which are not in the ideal calculation.

The average electron energy measured by the Langmuir probe is approximately 1.5 eV over the extraction grid area (and is not affected by the addition of argon). This is optimal for H^- production by volume processes and for the reduction in H^- loss by electron collisions [12]. However, the average energy steeply increases at low values

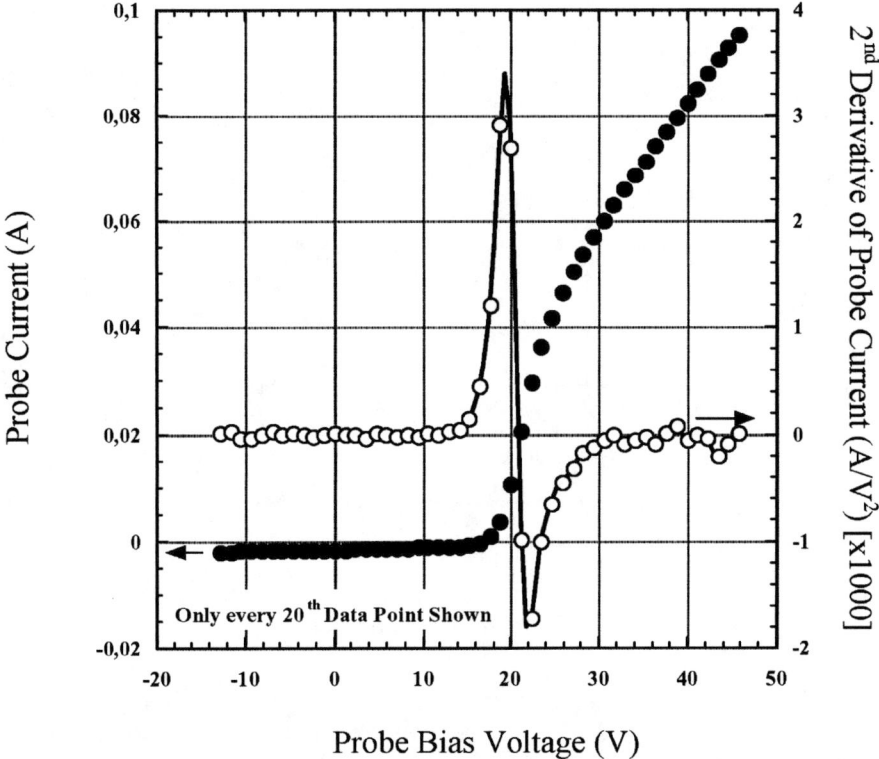

Figure 9. Shown is the Langmuir probe current as a function of applied probe bias. The data has been smoothed via a software "Blackman Filter." Also shown is the 2nd derivative of the probe current, the solid curve is a cubic spline fit to the data generated by the graphing software. The data was collected from a plasma of H_2 and Ar at 0.48 Pa and 120 kW of applied RF power.

of the probe position. In the case of the average electron energy it is possible that this is a measurement artefact brought about by the magnetic field.

The plasma potential, as normal, shows only an affect from the addition of argon to the plasma and is otherwise uniform across the surface of the source. A small increase is observed with increasing RF power: $\Delta U_{pl} = 0.7$ V when the RF power is increased from 60 to 110 kW. Although not shown in this paper, measurements have been conducted under identical conditions without Argon addition and the results also show this increase in U_{pl} with RF power.

As the results over the extraction area of the plasma grid are reasonably uniform for all parameters it is possible to use the averaged central values (the three data points near the 15 cm scan position) and plot these as a function of RF power. In the case of the average electron energy there is no significant power dependence observed and no effect from the addition of argon to the plasma (see Fig. 13). For the electron density a substantial difference in the behaviour between an argon-seeded plasma and a pure hydrogen plasma is observed (see Fig. 13 and 14). In Fig 14 it can

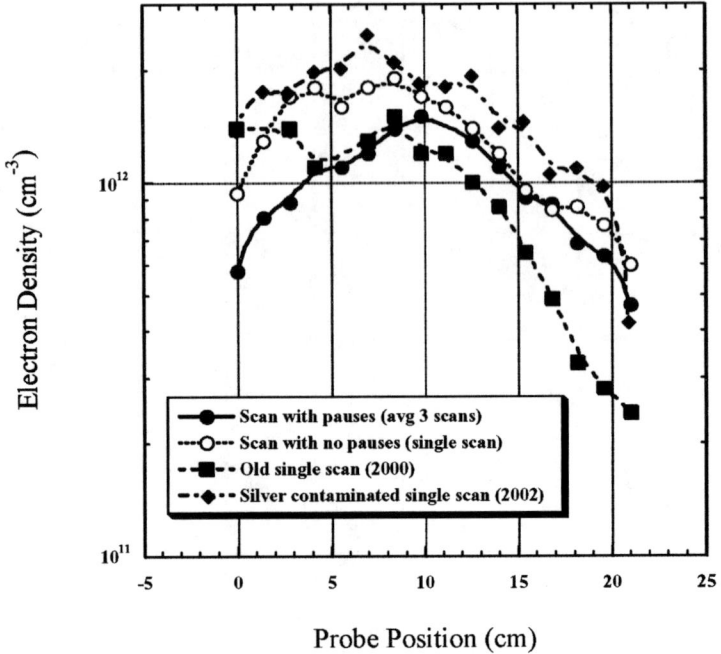

Figure 10. Shown above a Langmuir probe scan in the z-direction showing the effects on electron density of different methods of performing the scan with or without pausing for measurement. Also shown is a scan performed on the source when it was contaminated with silver. An older single scan is given as a reference. The discharge conditions for all scans were a pure hydrogen discharge at 0.5 Pa and 60 kW applied RF power. The plasma grid is located 28.4 mm from the 21 cm probe position.

be seen that the behaviour of the electron density mimics the power dependence of the extracted H⁻ current density. It is not possible to conduct Langmuir probe measurements simultaneously with beam extraction, however. So it possible that the plasma during extraction is not the same as when the Langmuir probe measurements were carried out. Under the assumption that extraction does not change the plasma, in the case of no argon addition, the pure hydrogen plasma rapidly reaches a saturation value of electron density and increased RF power has little effect. In the case of an argon-seeded plasma, however; the electron density increases largely linearly up to the maximum power used. As the electron density is critical to volume production mechanisms this is a qualitative explanation for the enhanced H⁻ observed under argon addition and in addition it explains the higher co-extracted electron current also always observed when argon is added to the plasma.

The results for the plasma potential are also of extreme interest. The plasma potential with argon is ~2 V lower than that of a pure hydrogen plasma (shown in Fig. 13). Both curves have similar slopes and the reduction appears constant and independent of applied RF power. In the past, in z-profile scans, a reduction in U_{pl} has

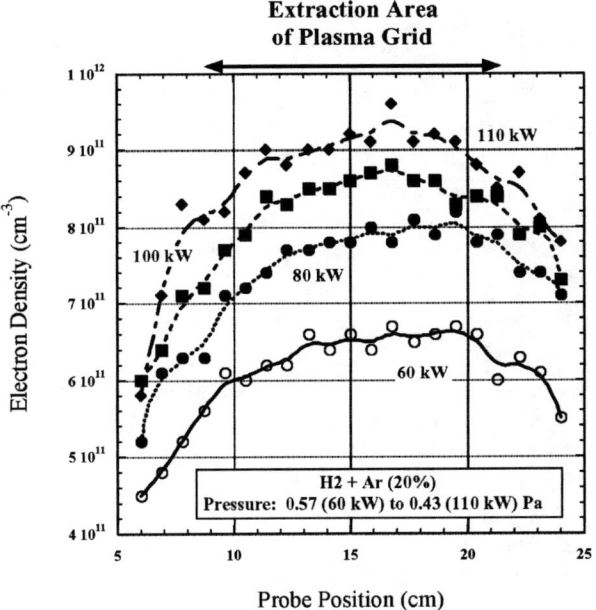

Figure 11. The electron density of the source as measured by the horizontal Langmuir probe for a mixture of hydrogen and 20% argon and different applied RF powers.

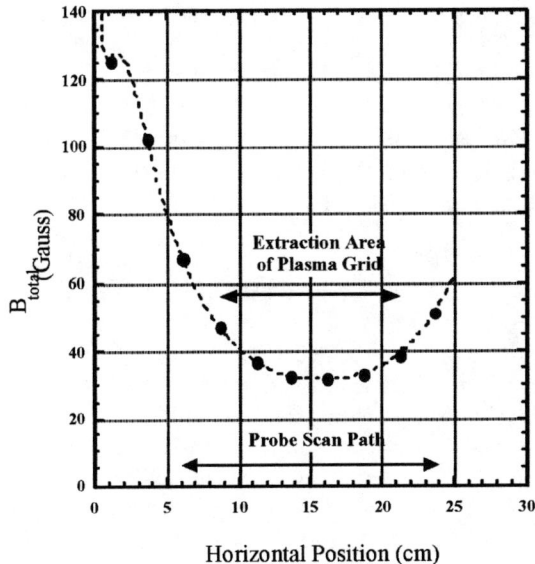

Figure 12. A calculation of the magnetic field of the magnetic filter along the path scanned by the horizontal Langmuir probe.

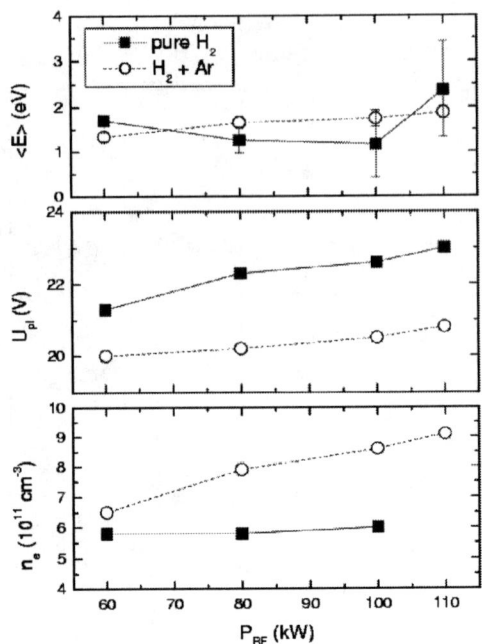

Figure 13. The central value of the plasma parameters in the middle of the magnetic filter shown as a function of RF power and the presence of argon in the plasma. The source pressure was 0.5 Pa.

also been observed [6] due to the addition of argon but it was not this pronounced being only 0.5 V. The effect is also of interest in that it is opposite to what one would expect based on ambi-polar diffusion: U_{pl} with argon is lower than the U_{pl} without. The reduction of the plasma potential has an impact on the extraction of H⁻ from the source as it reduces the potential hill the ions face.

V. Silver Contamination Effects

The first grid material to be tested in a new series of PG material experiments was a silver plasma grid and Faraday shield. It was installed in November 2001 and several weeks of experiments were performed with it. As time progressed source performance slowly degraded until by January 2002 it had reached the state where it was totally unacceptable. The source was opened at this time and the entire inner surface was found to be silver covered. Although, sputtering was the obvious cause of this, initially this seemed improbably due to the fact silver is heavier than copper and no such degree of sputtering had ever been observed with the copper Faraday shield.

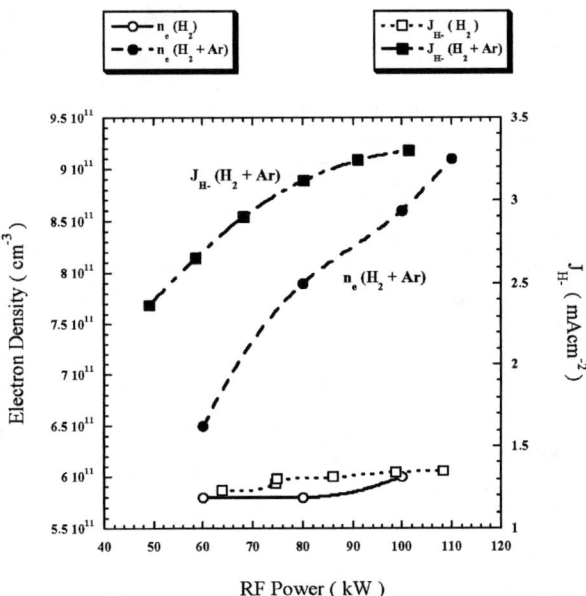

Figure 14. A comparison between the central electron density as a function of applied RF power for a pure hydrogen and a hydrogen-argon plasma. Also shown is the H⁻ current density for discharge parameters of 0.5 Pa and 20% Ar.

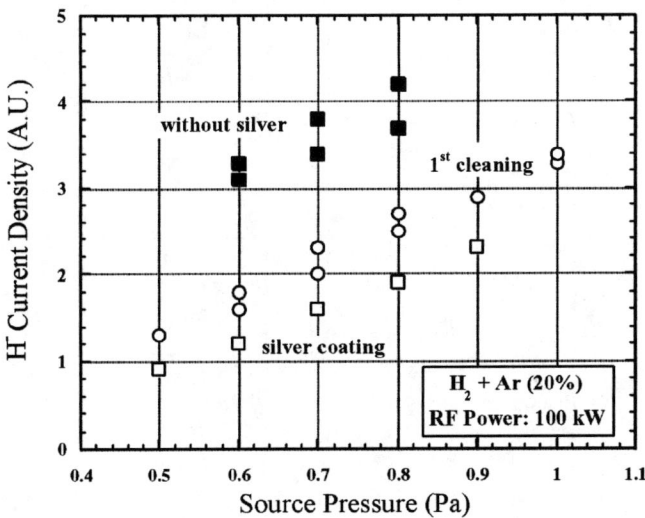

Figure 15. The performance of the Type VI ion source as a function of source pressure for hydrogen-argon discharges at 100 kW applied RF power showing the effect of silver contamination. Shot to shot variation in yield is shown by the different densities for otherwise identical conditions.

Further investigation into sputtering yields of silver under both hydrogen and argon ion bombardment revealed that the threshold for sputtering of silver by Ar^+ ions is only 17.4 eV as compared to the threshold for the sputtering of copper which is 27 eV[13]. As 17.4 eV is less than the measured plasma potential ($\approx$30 V) by a substantial amount the yield is appreciable. Complicating matters is that the self-sputtering of Ag^+ on silver has also a high yield (0.2 per incident ion in our energy range) and low threshold.

It was decided that the source would be returned to its normal configuration of Mo plasma grid and copper Faraday shield. After cleaning, operation commenced and although source performance improved (see Fig. 15) it still did not achieve nominal values from the past. The most disturbing observation was that the H^- current density, J_{H-}, was saturating at extremely low RF powers and even decreasing as the power was increased (see Fig. 16). There was no alternative to again dismounting the source. It was a considerable surprise to find that, again, the inner surfaces of the source were covered in silver. As no obvious source of the silver existed the only option was to completely disassemble the source and rigorously clean all surfaces.

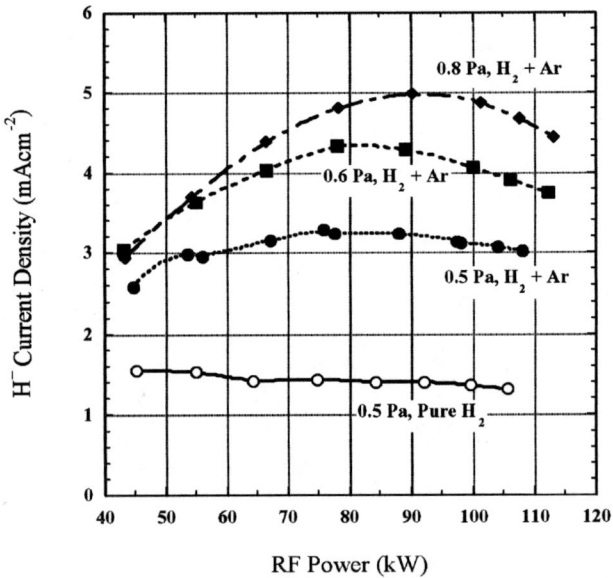

Figure 16. Power and pressure scans for a silver contaminated discharge. The early saturation of J_{H-} with silver contamination is clearly visible.

VI. Future Plans

In the near future the "large area grid"[16] will be installed on BATMAN and this will allow for a direct comparison of the low penetration magnetic electron suppression field of that grid with the current high penetration magnetic configuration. At the same time a system to selectively mask off portions of the grid will be installed on the current BATMAN grid. This will allow investigation on the effect of lowering systematically the extraction area plus will allow for investigations of the effect each aperture has on its neighbours by masking adjacent holes.

More spectroscopic measurements in collaboration with the TU Augsburg are planned and with the horizontal scanning probe functioning the analysis should proceed much faster as accurate measurements along the spectroscopic line of sight can be made between light collection shots.

The horizontal scanning probe in its opening debut has performed above expectations and significantly more experiments can be carried out in the future exploring the changes in the plasma due to caesium, heated grids, different magnetic configurations etc. The potential of mapping the plasma surface using the other ports also exists.

Although no plans exist to run with a silver grid in the future, the experience of excessive sputtering from the Faraday shield has led to a critical review of possible materials for the shield. It is clear that the most important criteria for the choice of Faraday shield material must be both low sputtering yield and high sputtering threshold for Argon ion bombardment. Initial results from the Augsburg group found no indication of the presence of Ar^{++} ions in the plasma, which is welcome news as the doubly charged ions have a significant sputtering yield and are with the plasma potential in the source above the sputtering threshold for all materials. It has been decided to coat a Faraday shield with tungsten and use this in future grid material tests. Although the possibility of contamination of the plasma grid with tungsten exists this is less of a concern than again contaminating the source with a metal which appears to poison H⁻ production and being required to invest a great deal of time and effort in removing it. Of considerable concern, in terms of sputtering from the Faraday shield, is the possibility of enhanced ion temperatures due to plasma compression from the time varying RF induced magnetic field [14]. Experiments are planned to measure the magnitude of this effect by measuring the magnetic field in the driver region.

VII. Conclusion

This paper summarises the results of recent investigations into both beam extraction and the extraction region of the Type VI ion source. For the first time it has been possible to study the profile of the beam on the calorimeter and not just the extracted beam current density as a function of source parameters. The beam width has found to vary qualitatively with current density as had been in the past calculated. An unexplained, as yet, change in the ratio of the horizontal to vertical beam width has been observed. The extraction voltage necessary to operate the source in the plasma limited regime has been determined to be 3 kV and excepting at extremely low current densities the source follows the expected dependence on extraction voltage in the space charge limited regime. All of the quoted current densities are determined by evaluation of calorimeter data as in the past. The perveance optimum of the extraction system has been determined to be 2.2×10^{-6} A/V$^{3/2}$ some 85% of the theoretical value for this geometry. Experiments made under conditions where the plasma grid was biased show for the first time an enhanced extracted current density as was observed by other groups but also show differences from the past experiments at Garching due to silver contamination.

The silver contamination of the source was an unfortunate diversion of experimental effort but even so can be said to have produced two important insights into source operation: that the sputtering properties of the material of the Faraday screen under argon ion bombardment are of critical importance, and that clearly the source is sensitive to surface effects. The latter is a strong indication that future investigations into "surface assisted volume processes" [15] may be fruitful.

A new scanning Langmuir probe has been installed in the diagnostic magnetic filter flange and has allowed up to study the plasma parameters near the extraction region and more importantly under typical operating conditions of the source. A clear correlation between the extracted ion current density and the electron density as a function of applied RF power was found. This is the first time so clear a correlation between the addition of argon, a change in plasma parameters, and the change in extracted ion current density has been observed. The plasma potential is also observed to decrease by some 2 V when argon is added to the plasma. This observation confirms measurements in the past but no explanation for this apparent deviation from the expected effect (an increase in plasma potential due to ambi-polar diffusion effects) exists.

The continual evolution of the data acquisition system on the testbed is very positive. The continued development of improved diagnostics such as the new horizontal scanning Langmuir probe and the new calorimeter while not enhancing the yield of the source directly should allow for a much better understanding of the processes involved in both production and extraction of negative hydrogen ions. And this understanding is critical if the ITER goals are to be met. The new collaborative effort on spectroscopy with the TU Augsburg has already provided valuable information and it is expected that the next experimental campaign will be even more profitable.

REFERENCES

1. Vollmer, O., *et al.*, *Fusion Engineering and Design*, **56-57** 465-470 (2001).
2. Vollmer, O., *et al.*, in *Fusion Technology 1992*, C. Ferro, M. Gasparotta, H. Knoefpfel (eds), Elsevier Science Publishing (1993) 1106-1110.
3. Hu, C., "H⁻ Stripping Loss in the Tank Zone", Internal TE-AGNI report (2001) unpublished.

4. Fantz, Ursel, TU Augsburg, private communication and Proceedings of the 3rd Workshop on Inductively Coupled Plasmas, Bochum, Germany (2002) unpublished.
5. Scheubert, P., *et al.*, *J. of App. Phys.*, **90**(2) 587-598 (2001).
6. McNeely, P., *et al.*, *Fusion Engineering and Design*, **56-57** 493-498 (2001).
7. Franzen, P., *et al.*, ITER NBI Review Meeting, Naka, Japan (2000).
8. Trainham, R., et al., in *8th International Symposium on the Production and Neutralisation of Negative Ions and Beams*, C. Jacquot (ed.), Giens, France (1997) 105-112.
9. Trainham, R., CEA Cadarache, private communication.
10. Whealton, J.H., *et al.*, *Rev. Sci. Instr.*, **71**(2) 939-942 (2000).
11. McNeely, P., and W. Kraus, Proceedings of the May CCNB Meeting, Cadarache, France (2001), unpublished.
12. Bacal, M., *Physica Scripta*, **T2**/2 467-478 (1982).
13. Eckstein, W., IPP Report 9/82.
14. Wilhelm, R., submitted to *Phys. Plasmas*.
15. Bandyopadhyay, M., Internal TE-AGNI report (2002) unpublished.
16. Heinemann, B., *et al.*, in *20th Symposium on Fusion Technology*, B. Beaumont, P. Libeyre, B. de Gentile, and G. Tonon (eds.), Marseille, France (1998) 433-436.

DIAGNOSTICS

Studies on a Magnetron Source

D. P. Moehs

Fermi National Accelerator Laboratory[a], PO Box 500, Batavia, Illinois 60510 USA

Abstract. In working toward a less noisy H⁻ magnetron source, a new diagnostic tool consisting of a scannable Faraday Cup Array has been developed and installed on the Fermilab H⁻ ion source test bench. The array consists of 14 identical Faraday cups with 1 mm diameter entrance holes spaced 6.4 mm apart. H⁻ beam currents of 50 to 80 mA with a radius of roughly 3 to 4 cm at the collector are routinely produced by the magnetron. Local current density measurements made as a function of acceleration voltage, sample time during the pulse, and background pressure reveal space charge compensation and beam self-focusing. An optical means of measuring simultaneously the local current density and local phase space parameters is also being investigated. These measurements should shed further light on how the current and beam divergence is distributed over the beam cross section.

INTRODUCTION

Future demands for protons at Fermilab have lead to the consideration of a new proton driver[1] with a front end consisting of a new H⁻ source and a multi-MeV RFQ. Currently, RF sources appear to be the source of choice based on their potential for high brightness although their source lifetime is still in question[2]. Better antenna coatings do appear to be improving their lifetime and an external antenna has already demonstrated 7000 hrs of operation. With additional R&D magnetrons, which have not been studied significantly since the 80's, may yet produce beams of comparable brightness and they already hold the combined record of intensity and longevity. Negative ion temperatures are difficult to reduce in the magnetron when they are run in the gas-starved mode since the cathode, upon which most of the extracted H⁻ ions are produced, directly faces the extraction aperture. Other ways of improving the brightness may be found by improving the extraction system or reducing source noise. In which case, the selection of which source to use may remain a matter of personal choice.

Immediate demand for brighter Linac and Booster beams at Fermilab has opened up the opportunity to carry out some new magnetron R&D on the Fermilab test bench. Traditional ways of increasing the current such as increasing the extraction voltage have been considered but limited space, sparking of the extractor and the inability to get sufficient flux to the 90 degree focusing magnet has limited the extraction voltage to 23 kV. Because of these constraints, other avenues for increasing the ion source

---

a Work Supported by University Research Associates Inc., contract number DE-AC02-76CH03000

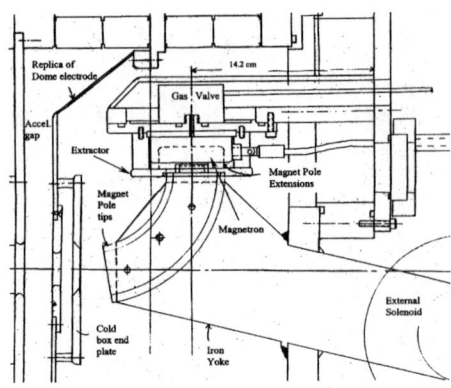

Figure 1. Side view of the Fermilab test bench showing the H⁻ Magnetron source, the magnetic poles and the extractor.

performance are being considered. Figure 1 shows a sketch of the test bench H⁻ magnetron and accelerating electrodes. A slit extraction aperture 1 mm × 10 mm is used. Detailed descriptions of the Fermilab set up and the typical operating parameters can be found in Ref. 3, 4 and 5. Present diagnostic instrumentation on the test bench after the acceleration gap, include a beam current toroid, 10 cm diameter Faraday cup, quartz view plate, pepper pot and XX' slit emittance measuring apparatus. To these diagnostic tools a scannable Faraday cup array with 1 mm resolution has been added allowing measurements of the local current density.

SOURCE MODIFICATIONS

One of the first efforts to produce higher currents with the Fermilab H⁻ magnetron has been to replace the anti-symmetric grooved cathode with a symmetric one. Alessi and Sluyters previously reported increased H⁻ currents from a magnetron with a backside anode-cathode spacing of 5 mm[6]. Although grooving the cathode all the way around increases the source volume slightly the pressure necessary to maintain the discharge has remained unchanged. This is important because high background pressure creates problems when dealing with ion pumps on the HV columns. In these tests, changing the cathode itself did not immediately lead to higher currents. However, when the magnetic pole extensions, originally designed to increase the B-field across the source, were removed there was roughly a 10 % gain in H⁻ current at an extraction voltage of 21 kV. Iteration of cathode type with and without the pole extensions verified that the increase in H⁻ current was associated with this combined change. Measurement of the H⁻ current as a function of arc current for different extraction voltages is shown in Fig. 2 for the original, anti-symmetric, and new symmetric source configuration. One feature associated with this current increase appears to be a smoothing out of the transition from the arc limited to the space charge limited region of operation. Since these currents were measured after the low energy transport system this smoothing may be associated more with transportation effects

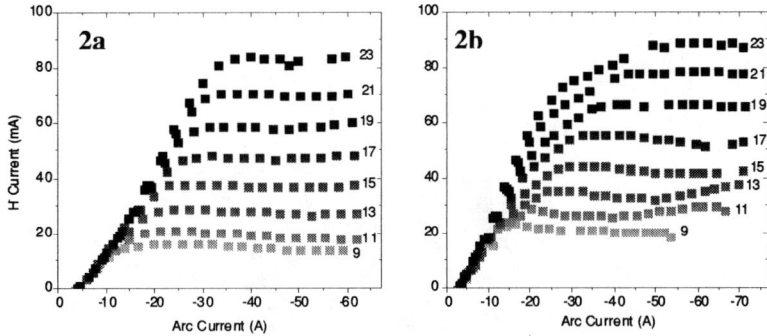

Figure 2. H⁻ current plotted as a function of arc current and extraction voltage for the anti-symmetric (2a) and symmetric (2b) cathode configurations. Grayscaling has been used to visually separate out the different extraction voltages, which are listed along the right side of the plot.

rather than source changes. In addition, it appears that the source is less noisy when using a symmetrically grooved cathode but further investigation is necessary to quantify this observation.

Recently this new source configuration was installed in one of the Linac preaccelerators. After standard start-up procedures, the current at the end of the 750 keV column for a 21 kV extraction voltage was 90 mA. By comparison, until recently the Linac was always tuned for peak currents in order to maximize the amount of charge injection into the Booster. In March and April of 2001 the extraction voltage on the same preaccelerator was near 21 kV. During that time, the average current at the end of the HV column was around 80 mA and the peak current was 87 mA.

CURRENT DENSITY MEASUREMENTS

Current Density Measurements have been carried out using a Faraday cup array consisting of 14 identical but independent Faraday cups mounted vertically on the end of a motorized linear feed through. Centered in front of each cup is a 1 mm diameter entrance hole spaced 6.4 mm apart. The 5 mm inner diameter of each cup was drilled 22 mm deep reducing the need for a repeller grid. Differences between the Faraday cup signals within the Faraday array were studied by mounting the array horizontally such that each cup would pass through the same part of the beam. Comparing the peak signals from each Faraday cup revealed a ±5 % variation in intensity. This data was averaged over many beam pulses thereby reducing the influence of pulse-to-pulse fluctuations in the beam, assumed to be small, and source noise, which creates current fluctuations of ±5 % across the 80 ms peak of the beam pulse.

Rough comparison to the large Faraday cup was made based on the average current density for the entire beam. The total beam spot size at the Faraday array was estimated from the endpoints of the array data. Over a wide range of operating conditions the average current density measured by the array was 5 to 25 percent higher than that of the large Faraday cup. Some of this discrepancy results from a

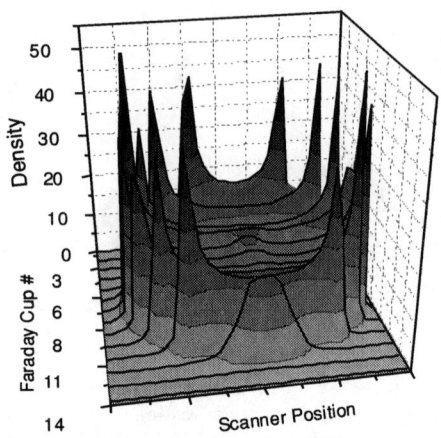

Figure 3. A current density measurement displaying electrostatic focusing of the beam on the edges with limited space charge neutralization appearing in the center. The density, echoed in the grayscaling, is given in $\mu A/mm^2$. For this data set, the extraction voltage was set near 21.5 kV while the acceleration voltage was close to 0.

poor estimate of the beam spot size, which is limited to the cup spacing in the vertical direction. In addition, secondary electrons created at the entrance apertures may also be contributing. Positive biasing of individual Faraday cups in the array produced increased signals indicating that a repeller grid may be necessary. Better beam spot estimates and studies of the systematic effects of secondary electrons are underway.

Individual current measurements are averaged over many beam pulses as the Faraday array moves through the beam. A complete local current density plot is composed of data from 20,000 or more beam pulses. The scanner position is incremented in roughly 0.1 mm steps while the resolution in Y is limited by the cup spacing. The linearity of motion in the X direction is excellent over the macroscopic range but deviations of up to 0.5 mm exist step to step and thus a scale is used. Improvements in the motion control program to be made this summer should rectify this situation. One of the strong advantages of this system is that the current density can be sampled at different times during the pulse. In this way, changes occurring during the pulse such as space charge neutralization can be studied. Three-dimensional reconstruction of the data reveals distinct signature differences between electrostatic focusing, affecting the outer part of the beam, space charge compensation, affecting the internal part of the beam, and beam scrapping. Figure 3 shows the first two of these effects.

Throughout this study there has been increasing evidence that scrapping of the beam on the 90-degree focusing magnet has been affecting transport efficiency. Cross-sectional beam images observed on a quartz plate and in current density measurements show clear deviations from the roughly circular beam that is expected after space charge blow up. Discoloration of the bending magnet near the exit plane has been observed for years and in the past it was always assumed that scrapped beam

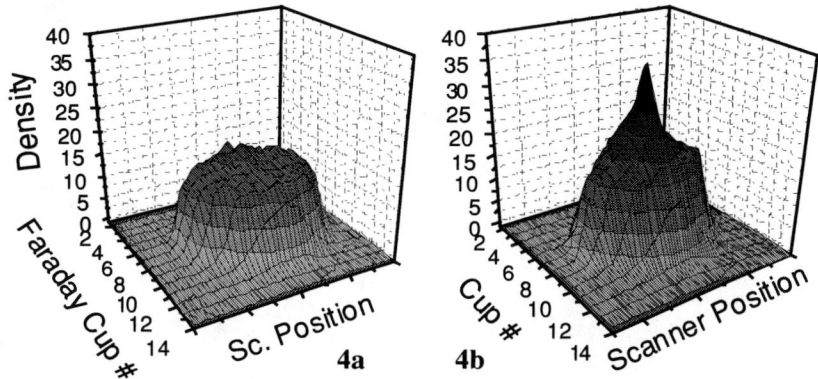

Figure 4. Local current density plots showing the affect of adding nitrogen as a background gas. In figure 4a the base operating pressure was 2.7×10^{-5} torr and the local current density is roughly flat across the top. In figure 4b the background pressure was increased with nitrogen to 3×10^{-5} torr enhancing space charge neutralization and causes a sharp central peak to form.

would fall outside of the accelerator acceptance. This idea is being questioned now due to the appearance of vertical asymmetries in beam spot images. A potential means of overcoming the limited space of the magnet would be to improve the space charge neutralization in this region. Fig. 4 shows the affect on the local current density when a small amount of nitrogen is added as background gas. In both cases a total current of 65 mA was measured for an extraction voltage of 21 kV and an acceleration voltage of 30 kV, applied at the acceleration gap. The addition of nitrogen reduced the overall beam spot size by roughly 35% and appears to focusing the beam, increasing the central current density by a factor of two. Drawbacks to this technique include increased gas loading of the ion pumps evacuating the HV column and the creation of beam plasma instabilities.

PRESENT STATUS AND FUTURE PLANS

With the 10% gain in H⁻ current from the source and new injection practices used to transfer beam from the Linac into Booster[7] an effective H⁻ intensity overhead of 30% has been achieved. Taking advantage of this overhead, a reduction in the width of the extraction aperture may now be possible in order to directly reduce the source emittance. It is expected that such a change will also allow the source to operate at higher pressures, which should lead to reduced beam noise and better emittance.

Plans are being made to optically measure the local and global phase space and current density. This should shed further light on how the current and beam divergence is distributed over the beam cross-section as well as provide a better understand of how changes to the source effect the beam quality. Work is also beginning on computer simulations of the extraction region in order to gain additional insight as to how changing the slit width will affect the emittance. A two-electrode extraction system is also being considered to help reduce sputtering of the emission slit and cathode by positive ions.

ACKNOWLEDGEMENTS

The author would like to thank Charles Schmidt for passing on his expertise with the Fermilab magnetron, Jim Wendt and Ray Hern for helping to revive the ion source test bench and Vadim Dudikov for many stimulating conversations.

REFERENCES

1. G.W. Foster, W. Chou, E. Malamud, Fermilab-TM-2169 (2002)
2. J. Peters, *Rev. Sci. Instrum.* **71**, 1069-1074 (2000)
3. C. W. Schmidt, 2nd Symp. on the Production and Neutralization of Negative Hydrogen Ions and Beams, BNL, 189-191 (1980)
4. C. W. Schmidt and C. D. Curtis, 4th Symp. on the Production and Neutralization of Negative Hydrogen Ions and Beams, BNL, 425-429 (1986)
5. C. W. Schmidt, LINAC Conf. Proc., LANL, 259-263 (1990)
6. J.G. Alessi and Th. Sluyters, *Rev. Sci. Instrum.* **51**, 1630-1633 (1980)
7. D. P. Moehs, L. J. Allen, E. S. McCrory, C. W. Schmidt, R. C. Webber and R. E. Tomlin, submitted for publication, LINAC Conf. Proc., Korea (2002)

Status Report of the Frankfurt H⁻-Test LEBT Including a Non-destructive Emittance Measurement Device*

C.Gabor**, A.Jakob, O.Meusel, J.Schäfer, A.Klomp, F.Santić,
J. Pozimski, H.Klein, U. Ratzinger

Institut für Angewandte Physik, Goethe-Universität, 60054 Frankfurt, Germany

Abstract: For high power proton accelerators like SNS, ESS or the planned neutrino factory (CERN), negative ions are preferred because they offer charge exchange injection into the accumulation rings (non Liouvillian stacking). The low energy beam emittance is a key parameter in order to avoid emittance growth and particle losses in the high-energy sections. Conventional destructive emittance measurement methods like slit-harp systems are restricted for high power ion beams by the interaction of the ion beam with e.g. slit or harp. Therefore a non-destructive emittance measurement has several technical and physical advantages. To study the transport of high perveance beams of negative ions, a *Low Energy Beam Transport* (LEBT) section is under construction. The study of non destructive emittance measurement devices is one major subject of the test bench. For negative ions -especially H⁻-ions- *photo*detachment can be applied for a non-destructive *e*mittance *m*easurement *i*nstrument (PD-EMI). The paper will present the status of that emittance diagnostic and of the test bench.

INTRODUCTION

The proof of principle to use photo neutralization for emittance measurements at energies higher than 700 MeV has been demonstrated earlier. [1,2] The outer electron of the H⁻-ion has an ionisation potential of 0.75 eV and can be effectivly removed by photons of a wavelength below 1500 nm. The peak cross section of $\sigma = 4.0 \ast 10^{-17}$ cm² [3,4] is reached for a wavelength of 830 nm. The basic idea of the diagnostic device is to scan the laser through the H⁻-beam within a magnetic dipole field (see figure 1). The neutrals produced by photodetachment will be separated from the negative ions

---

*supported partly by EU contract number HPRI-CT-2001-50021

**email : c.gabor@iap.uni-frankfurt.de; phone : 0049-69 798 23475 and from the other neutrals produced by the residual gas and their will be distribution measured.

This method has some advantages: No mechanical parts like slits must be driven through the beam, no secondary particles are produced, a defined degree of space charge exists and only a small portion of H⁻-beam will be neutralized by the photons. The produced H⁰-beam can be analysed without disturbing the ion beam. Further features are a good resolution in time and phase space. Therefore a PD-EMI seems to be suited for online measurements of high brilliant ion beams (e.g. near focus).

EXPERIMENTAL SETUP

To investigate several aspects [5] of low energy beam transport of H⁻ ions, including new non destructive diagnostic techniques a *Low Energy Beam Transport* (LEBT) test bench is under construction. It will be based in the beginning on a caesium-free, low current ion source [6,7] and on a double solenoid LEBT (max. field B=0.8 T). For high transmission a vacuum chamber with a pinhole ($\emptyset \approx \frac{3}{2} d_{BEAM}$) will be installed after the ion source. It subdivides the test bench into two vacuum sections (differential pumping). This will reduce the gas flow into the LEBT (p ≤ $5*10^{-7}$ hPa) and therefrom reduce particle losses by residual gas interactions. Additionally the residual gas pressure can be varied independently. This is necessary for the planned investigations on space charge compensation. The LEBT will be capable of transporting high perveant beams (up to 100keV and 140mA protons or H⁻). Furthermore the test bench will include conventional beam diagnostics like Faraday cups, an emittance scanner of the Allison type and residual gas ion energy spectrometers.

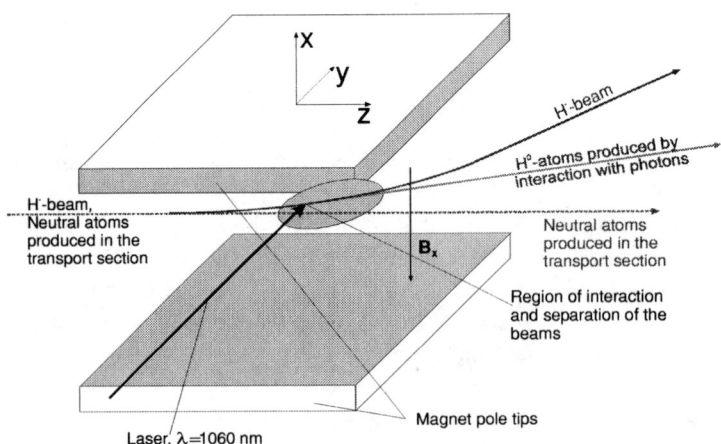

FIGURE 1. Principle of the planned photodetachment emittance measurement instrument (PD-EMI).

Downstream the second solenoid lens a new diagnostic chamber including the PD-EMI will be installed [8]. Most parts of the beam line already exist and are operational. At the beginning first beam measurements with helium and hydrogen have been carried out to calibrate the mass spectra measured behind of a 90° dipole magnet. The experimental set up is shown in Figure 2.a, a mass spectrum in Figure 2.b. The tests will continue to meet the design values (H$^-$, 2.3 mA, 6.5 keV) of the source. To improove the signal to noise ratio (see next chapter) in a next step it is intended to redesign the extraction system for higher extraction voltages (15-20kV) and to increase the ion current. The layout of the final vacuum tank of the test bench for beam diagnostics including the PD-EMI is in discussion.

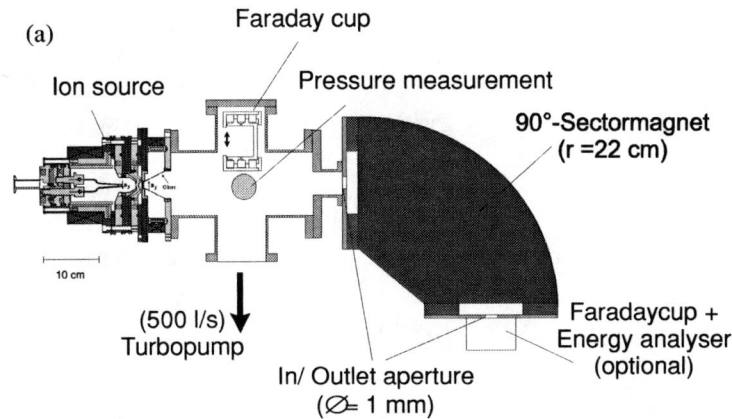

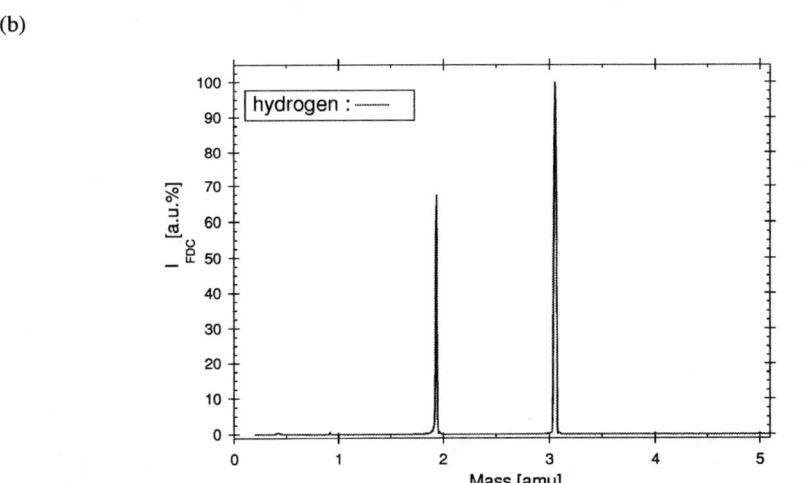

FIGURE 2. Schematic drawing of the experimental set-up of the magnetic spectrometer (a) and first results of mass spetra with hydrogen at positive extraction voltage (b): The ion source delivers mainly H_3^+ and H_2^+.

THE LASER DIAGNOSTIC DEVICE

For the right choice of the laser system used for photo detachment, the number of neutralized H⁻ ions have to be calculated. The production rate for photodetached H^0 is given by:

$$\dot{n_0} = \sigma * n_{H^-} * n_\gamma * v_{rel} \quad (1)$$

where n_{H^-}, n_γ are the densities of the H⁻-ions and photons; $\dot{n_0}$ is the production rate of the neutralized atoms [m⁻³s⁻¹] and at low energy transport normaly $v_{rel} = |\vec{c}|$. For low laser intensities (that means $n_H = $ const.) the total number of produced H^0 is given by

$$N_0 = \dot{n_0} * \tau * V_{reac} \quad (2)$$

where τ the pulse duration of the laser beam and $V_{reac} = \pi r_{laser}^2 * 2 r_{ionbeam}$ the reaction volume.

At high laser pulse energies the production of neutralized atoms H_0 leads into saturation and is given approximately by (assumed that a cylindrical laser beam passes through the center of a cylindrical ion beam where $r_{laser} < r_{ionbeam}$) [1,2] :

$$N_0(E) = N_H \frac{\tau}{t_i} \left(1 - e^{-\sigma \frac{\lambda}{h c r_{laser}^2 \pi \tau} \frac{2 r_{laser}}{v_H} E} \right) \quad (3)$$

where E = the laser pulse enery and the ions spent $t_i \ll \tau$ (t_i is the time of illumination) in the laser beam. Both graphs are shown in Figure 3.

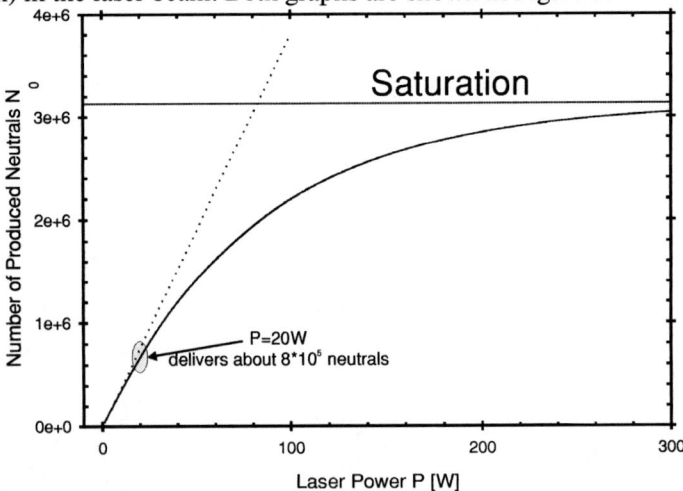

FIGURE 3. Evolution of the number of N_0 at the following parameters: 2mA H⁻, 20keV, d_{beam}=4cm, d_{laser}=1µm, τ=10µs, $\sigma(E)$=3.5*10⁻²¹m². These values chosen as the minimum case with respect to H_0-production. More realistic values for d_{laser}=250µm, τ=100µs deliver up to N_0= 3.6*10⁶.

In order to get a high degree of efficiency Nd:YAG lasers are a good choice. They deliver photons in the near infrared working at 75% of the peak cross section. For good spatial resolution the laser beam diameter must be small (d<0.5mm) throughout the diameter of the ion beam and must have a defined intensity profile. Therefore a high beam quality with small divergence angle and a preferably ideal gaussian beam profile M^2 are essential which results in the condition to apply a TEM_{00} mode only. To obtain this beam quality a laser resonator without additional optical elements has to be used. A Ytterbium Yb:YAG laser system with special pumping scheme meeting the beam quality requirements has been chosen (ELS [9]) and delivered. The beam power and quality tests have been successfully performed (M^2=1.01 at P_{max}=20.2W, d_{BEAM}≈1.5mm, λ=1030nm, beam divergence, full angle <0.5 mrad). For time resolved measurements a cavity-external acusto-optic modulator (AOM) working at 80 MHz was installed. Therefrom a time resolution below 1μs is available. The chosen technique of a DC Laser with external chopping gives high flexibility on exposure time.

Comparison Between Slit-Multiwire And PD-EMI

For the transport of a mono energetic ion beam it is necessary to know the 4 dimensional transversal phase space distribution $\rho(x, x´, y, y´)$. A common measurement device for emittances like the slit-harp system measures the positions (x, y) and angles (x´, y´) of the beam ions independently. The spatial position of the front slit determines the position of the ions. The rear harp measures the angular distribution (x´) of the trajectories passing the front slit (figure 4) and the result is the intensity distribution I (x, x´) in phase space. Due to the integration on the second transversal dimension this has to be done for both planes independently to get the full information about $\rho(x, x´, y, y´)$.

The PD-EMI works in a similar way. The laser beam scans like the slit perpendicular through the ion beam and produces a small number of neutrals H^0. In contrary to the harp in this case the fluorescent screen delivers a 2 dimensional image of the distribution of the neutralized ions. Therefrom additional information concerning the spatial and angular distribution of the second transverse dimension can be gained. Simulations with LINTRA [10] have been carried out and show after a short drift of z=10cm the influence of x´, y´on the x,y distribution. The simulation of the screen image has shown additionally the filamentation of the emittance.

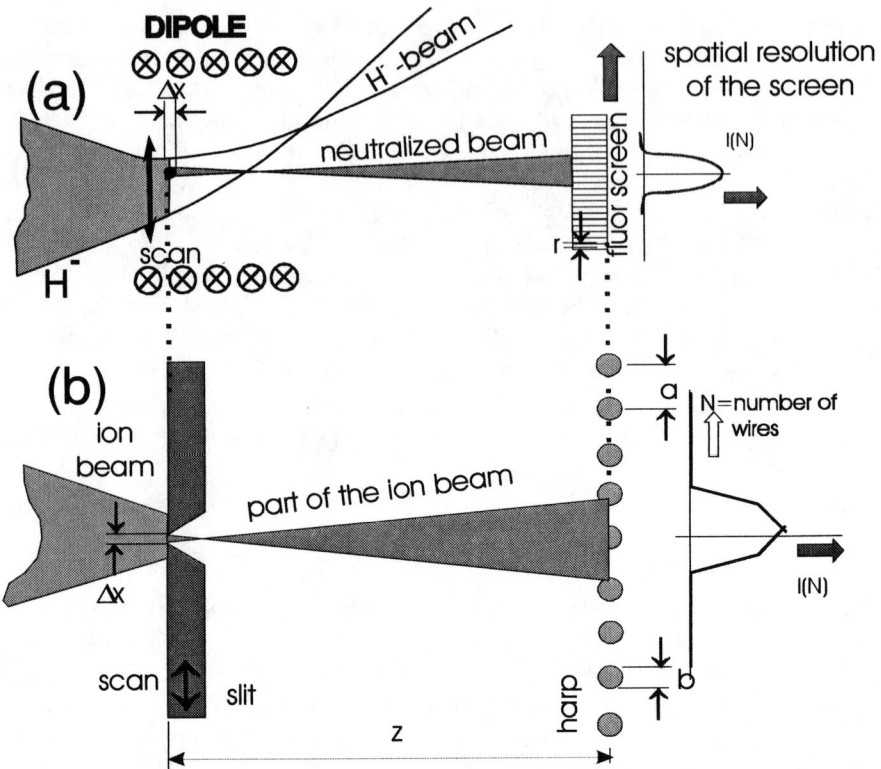

FIGURE 4. Comparison between PD-EMI (a) and slit-harp system. Δx is given by the laserbeam diameter resp. slit high and $\Delta x' = \arctan(r/z)$. A slit-harp emittance measurement device has an effective angle resolution of $\Delta x'_{eff} = \arctan(a/z)$ and a minimal $\Delta x'_{min} = \arctan(b/z)$.

OUTLOOK

The test bench will be completed in the new future. This includes the improvements of ion source, extraction system, differentially pumping vaccum system, electron dumping and the magnet chamber with dipole. The development of suitable optical elements to focus and to scan the laser beam is another point. Experiments to evaluate the performance of the PD-EMI in comparison with a conventional system are planned as well as investigations on the emittance growth under different conditions. The influence of the dipole field on beam transport and quality is another subject of investigations. A 2 dimensional scan with two laser beams is as well under consideration as the use of the PD-EMI near the ion source where the dipol field could be used for electron dumping too.

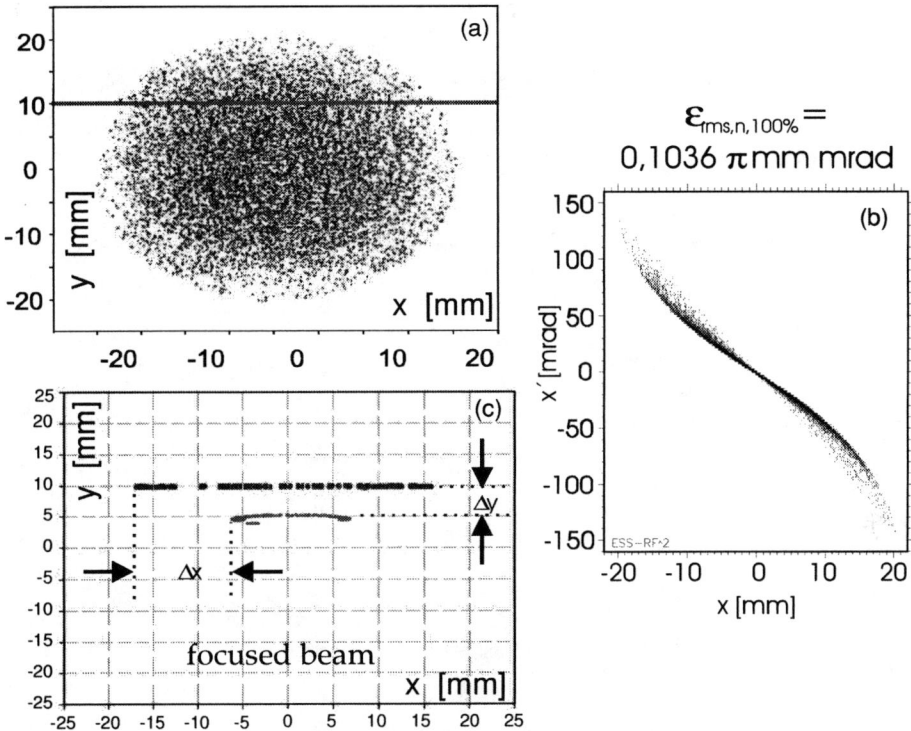

FIGURE 5. (a) shows the simulated xy distribution of the H⁻-beam with visualization of the laser beam. (b) shows the xx′ emittance at that position and (c) shows the original as well as the drifted distribution after 100mm. Due to a convergent beam, the image of the drifted distribution is smaller than at the position of neutralization. The deviation from a straight line is caused by the filamentation in phase space.

REFERENCES

1. Conolly, R.C., Johnson, K.F., Sandoval, D.P., Yuan, V., *Nucl. Instr. Meth.*, **A312,** 415-419, (1992).
2. Yuan, V.W., Conolly, R.C., Garcia, R.C., Johnson, K.F., Saadatmand, K., Sander, O.R., Sandoval, D.P., Shinas, M.A., *Nucl. Instr. Meth.*, **A329**, 381-392, (1993).
3. Ajmera, M.P., Chung, K.T., *Phys. Rev.* **A12**, 475-479, (1979).
4. Broad, J.T., Reinhardt, W.P. *Phys. Rev.* **A14**, 2159-2173, (1976).
5. Jakob, A., Gabor, C., Meusel, O., Pozimski, J., Klein, H., Ratzinger, U., these proceedings.
6. Jakob, A., Klein, H., Lakatos, A., Pozimski, J., Proc. 6th EPAC, Stockholm, 1400-1402, (1998).
7. Volk, K., Klein, H., Lakatos, A., Maaser, A., Weber, M., *Rev.Sci.Instrum.* **67**, 1039-1041 (1996).
8. Gabor, C., Jakob, A., Meusel, O., Pozimski, J., Schäfer, J., Ratzinger, U., *Rev. Sci. Instr.* **73**, No. 2 (2002).
9. ELS Elektronik Laser System GmbH, http://www.els.de
10. J.Pozimski, thesis, IAP, Johann Wolfgang Goethe-Universität (1997).

Diagnostics at the Frankfurt H⁻ - LEBT

A. Jakob, C. Gabor, O. Meusel, J. Pozimski, H. Klein, U. Ratzinger

Institut für Angewandte Phyisk, Johann Wolfgang Goethe-University, Robert-Mayer Str. 2-4, D-60054 Frankfurt am Main

Abstract. A magnetic Low-Energy Beam-Transport line for negatively-charged hydrogen ions was set up in Frankfurt, Germany. In order to study the space-charge compensation process several conventional beam diagnostic principles were already established. But to prevent compensated ion beams from distortions caused by conventional measurement devices, non-destructive methods are to be developed: Investigations were carried out to determine the emittance of the ion beam based on laser driven photon detachment of the H⁻-electron using a non-destructive method. This method is capable of determining the transverse phase space distribution with high time resolution and a large number of phase space points [1]. Additional non-destructive diagnostic tools are installed to study the temporal development of the compensation process. For time resolved measurements an optical device using CCD-cameras for beam profile measurements, a Residual Gas Energy Analyzer, and an Electron Energy Analyzer for beam potential measurements were installed. The working principles of the measurement devices as well as the whole test bench and planned activities are described.

1 INTRODUCTION

Positive and negative particles are produced by interaction of beam ions with residual gas. Due to accumulation of compensation particles (positive ions in case of negative ion beams, electrons in case of positive ion beams) the beam potential will decrease. In case of partly compensated beams, the rest gas component with a polarity identical to the beam ions is accelerated radial out by the remaining beam potential.

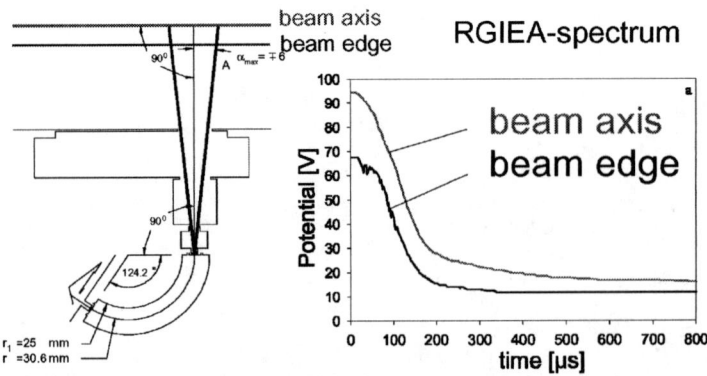

FIGURE 1. Schematic of the Energy Analyser, typical time resolved measured beam axis and beam edge potential (He⁺, 10keV, 2mA, $p_{RG}=2.7*10^{-5}$hPa).

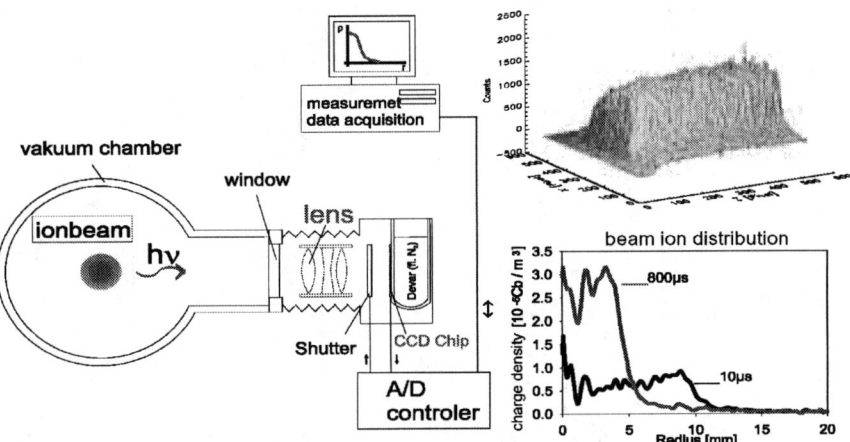

FIGURE 2. Scheme of the CCD-camera with intensifier and lens coupling system. Typical CCD-chip signals measured at positive ion beams (He$^+$, 2mA, 10keV, p_{RG}=2.7*10^{-5}hPa) and beam profiles for a partly compensated ion beam (10 µs after compensation start) and compensated state (800 µs a.c.s).

To study space-charge compensation processes, investigations of the beam potential and of the potential distribution are to be carried out. To study the time development of the beam potential a 127° Hugh Royanski-type Residual Gas Energy Analyser (RGEA) and an Electron Energy Analyzer (EEA) will be used.

Beam potential measurements of positive ion beams using a RGEA are well established. To allow single particle detection, the Energy Analyzer is equipped with a channeltron. Therefore, time resolved measurements of both beam axis and beam edge potential are possible. Typical beam potential measurements are shown in the right plot of Figure 1. The temporal development of the compensation process is indicated by the temporal development of the beam axis and beam edge potential. These measurements show that the compensation process is complete after approximately 400 µs in case of a 10keV, 2mA He$^+$ beam with a residual gas pressure of p_{RG} = 2.7*10^{-5} hPa in the diagnostic chamber.

As another non-destructive measurement device, a liquid N$_2$ cooled CCD-camera will be used (figure2). The CCD-camera is equipped with an intensifier to allow time-resolved measurements within a minimum gate time of 50 ns. The second plot in Figure 2 shows signals detected by the CCD-chip (512x512 pixel). The third plot in Figure 2 shows beam profiles at different compensation times extracted from measured data.

Investigation of effects due to beam asymmetries

Due to the magnetic filter field [2] of the H$^-$-source the extracted ion beam will be non cylindrically. Therefore two spectrometers and two CCD-cameras in an orthogonal arrangement (figure 3) will be installed to study effects due to non symmetrical beams [3].

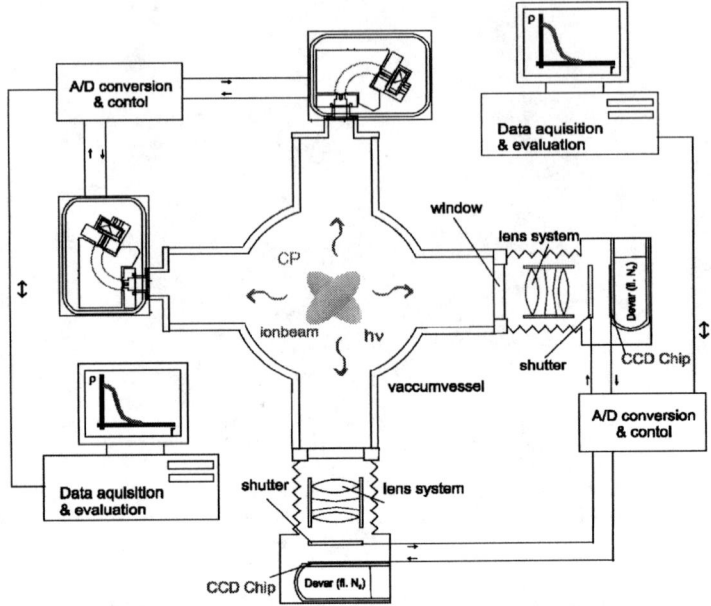

FIGURE 3. Diagnostic set up of two CCD-cameras and of two spectrometers in orthogonal arrangement.

The diagnostic system will be operated by a central control unit to allow time-synchronous measurements. By comparison of time-resolved measurements in both transverse directions, effects resulting from a change of the beam particle distribution will be studied.

Investigation of over compensation

An effect will be investigated, which can only be expected on negative ion beams: Due to the mass of the compensating ions and contrary to the compensation of positive ion beams, a state of overcompensation is expected. In this case, the compensation of positively charged residual gas ions can possibly exceed that of the H$^-$ - beam ions. The dependence on the beam pulse structure will be of special interest. To measure such effects the energy analyzer for electrons and ions will be used.

Laser driven emittance measurement device

As a further diagnostic tool for negative ions an online non-destructive emittance measurement system capable of being used for beams of diameters up to 40mm are being developed by C. Gabor [1]. The method is based on H$^-$-neutralization by laser detachment of H$^-$-electrons [4]. The measurement principles are shown in figures 4&5.

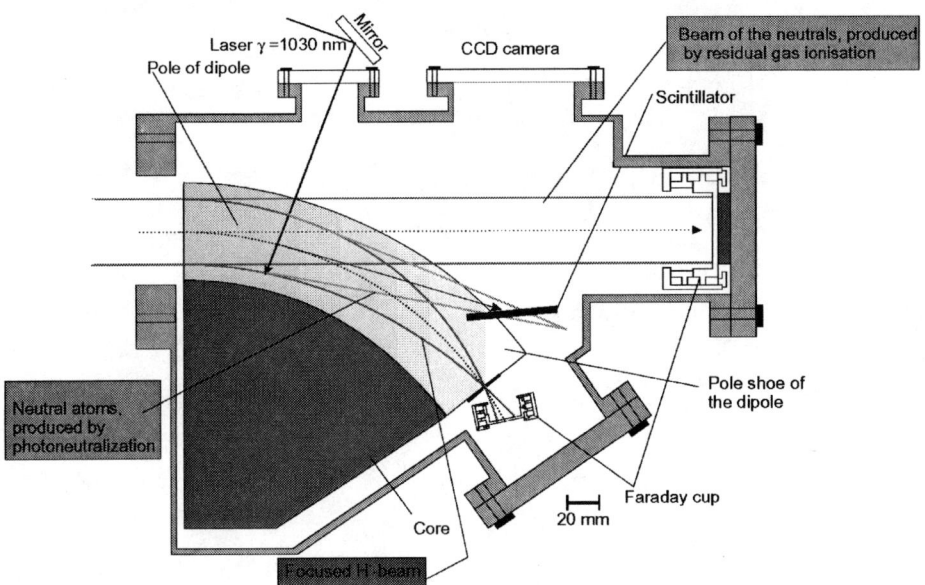

FIGURE 4. Schematic drawing of the diagnostic chamber and the dipole of the laser driven emittance measurement device.

The interaction of laser light and H⁻-ions produces a small number of neutral atoms and detached electrons. Neither the laser photons nor the recoiling photo-detached electrons transfer a significant momentum to the H-atoms. Therefore, it can be assumed, that the beam of neutralized ions has the same distribution in the 6-dimensional phase-space as the primary H⁻-beam.

The ion beam is separated magnetically from the neutral beam in a dipole field (Figure 4). The distribution of the photo-neutralized hydrogen atoms is measured using a CCD-camera by observing the light produced by the interaction of laser neutralized H-atoms with a scintillator.

Figure 5 shows the functional principle and the diagnostic set up. The transversal light distribution on the scintillator as a function of the intersection line between the laser and ion beam defines the divergence angle which finally results in the emittance of the ion beam. The spatial resolution is defined by the "virtual slit" of the laser diameter. The angular resolution is given by the drift length (at low gas pressure an undisturbed drift can be assumed), the scintillator and the CCD-camera.

The laser (λ = 1060 nm, 20 W) works in cw mode. For time-resolved measurements, an external **A**custo **O**ptic **M**odulator (AOM) is used, which diffracts the laser beam in a beam dump with a repetition rate up to 80 MHz. The time resolution Δt is defined by the measuring system (laser power, scintillator and CCD-camera). It is in the range of 10-500 μs and will be below typical beam pulse lengths of 1.2 ms (ESS, NIS).

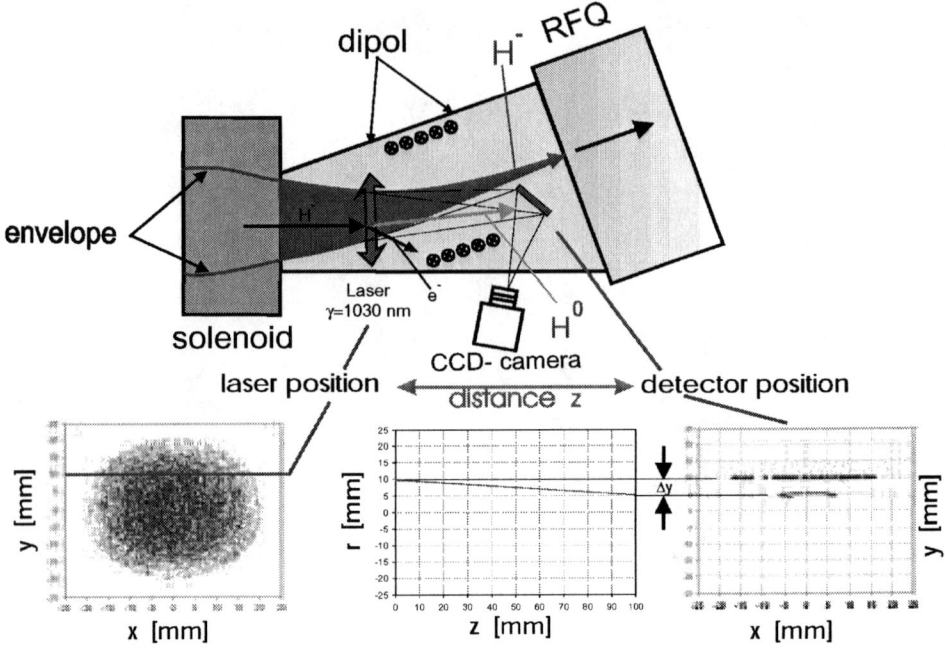

FIGURE 5. Schematic drawing of the diagnostic set up of the laser driven emittance measurement device.

2 EXPERIMENTAL SETUP

The LEBT section is designed to fulfill the requirements of the ESS and NIS project (>70mA, 35kV, 2.5ms, 50Hz). The experimental setup for the first phase of measurements is shown in figure 6.

The LEBT section consists of an ion source, an extraction system (triode, 65 kV), a first diagnostic chamber (Faraday cup), two solenoids, and a second diagnostic chamber with installed Residual Gas Ion Energy Analyzer (RGIEA), a liquid N2-cooled CCD-camera with installed intensifier, a conventional slit – slit emittance measurement device, and a Faraday cup.

Firstly, measurements will be done using proton beams to check the alignment of the LEBT system and of the diagnostics. These proton measurements will serve as reference for the investigations with H⁻-beams.

Secondly, the LEBT line will be tested in H⁻-operation using a low-current cesium-free ion source. For this purpose measurements of the beam potential (Electron Energy Analyzer EEA), of the beam profile (CCD-camera) and of the emittance (conventional emittance scanner) will be performed.

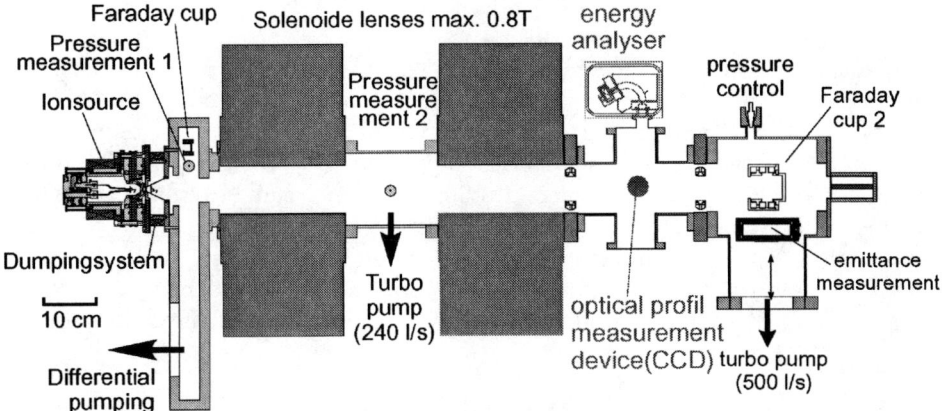

FIGURE 6. Schematic drawing of the LEBT section and diagnostics in first phase of measurements.

After testing the transport section the diagnostic chamber behind the second solenoid will be replaced by the laser diagnostic chamber in order to investigate the concept of a laser driven emittance diagnostic (figure 7) and to carry out non-destructive measurements of the beam emittance. Once this method has been demonstrated successfully in cw-operation, the laser driven emittance measurement device will be operated in time-resolved mode.

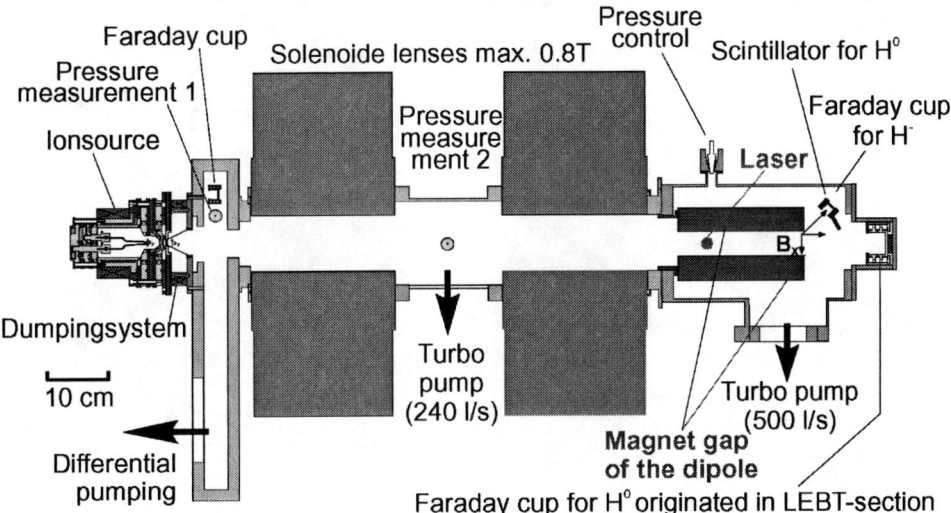

FIGURE 7. Schematic drawing of the LEBT section and of diagnostics in second phase of measurements.

3 SUMMARY

To study the compensation process for H⁻-beams and to prevent perturbations of the compensation process and, therefore, of the transport properties of the LEBT section, non-destructive, time-resolved measurement devices (residual gas ion energy analyzer, CCD-camera) are established. Measurements will be performed using particle energy analyzers (to detect residual gas ions and electrons) and CCD-cameras in an orthogonal arrangement to study effects on the compensation process and, therefore, on the beam transport due to the non cylindrical symmetry of the ion beam. To develop a non-destructive and time-resolved emittance scanner, the application of a laser driven emittance measurement device will be investigated.

REFERENCES

1. C. Gabor et al. Design of non-destructive emittance measurement device for H⁻ - beams, *Review of Scientific Instruments*, Volume 73, Nimber 2, February 2002
2. M. Bacal, G.W. Hamilton, E. Nicolopoulos and H.J. Doucet. *Phys. Lett.* 42, p. 1538, *J. Phys.* (Paris) 38, p 1399 (1977)
3. A. Jakob et al. „Investigation on temporal development of the Compensation Process of focused ion beams", *Proc. 7th European Particle Accelerator Conference (EPAC)*, Vienna, Austria, June 26-30, 2000 EPC, Austria Academy of Sciences Press, Vienna 2001,ISBN 3-7001-2931-9, OEAW CDR MN2, pp. 1741-1743
4. R. Connely. P. Cameron, J. Cupolo, D. Gassner, M. Grau, M. Kessemann, S. Peng and R. Sikora, Brookhaven National Lab., NY, USA, proceed. of *Beam Instrumentation Conference* April 2002

Accurate Estimation of the RMS Emittance from Single Current Amplifier Data[1]

Martin P. Stockli[a], R. F. Welton[a], R. Keller[b],
A. P. Letchford[c], R. W. Thomae[b], and J. W. G. Thomason[c]

[a]*Spallation Neutron Source, Oak Ridge National Laboratory, Oak Ridge, TN 37830, USA*[2]

[b]*E. O. Lawrence Berkeley Laboratory, Berkeley, CA 94720, USA*[2]

[c]*ISIS, Rutherford Appleton Laboratory, Chilton, Didcot, Oxon, OX11 0QX, UK*

Abstract: This paper presents the SCUBEEx rms emittance analysis, a self-consistent, unbiased elliptical exclusion method, which combines traditional data-reduction methods with statistical methods to obtain accurate estimates for the rms emittance. Rather than considering individual data, the method tracks the average current density outside a well-selected, variable boundary to separate the measured beam halo from the background. The average outside current density is assumed to be part of a uniform background and not part of the particle beam. Therefore the average outside current is subtracted from the data before evaluating the rms emittance within the boundary. As the boundary area is increased, the average outside current and the inside rms emittance form plateaus when all data containing part of the particle beam are inside the boundary. These plateaus mark the smallest acceptable exclusion boundary and provide unbiased estimates for the average background and the rms emittance. Small, trendless variations within the plateaus allow for determining the uncertainties of the estimates caused by variations of the measured background outside the smallest acceptable exclusion boundary. The robustness of the method is established with complementary variations of the exclusion boundary. This paper presents a detailed comparison between traditional data reduction methods and SCUBEEx by analyzing two complementary sets of emittance data obtained with a Lawrence Berkeley National Laboratory and an ISIS H⁻ ion source.

---

[1] This work is supported by the Office of Science, Office of Basic Energy Sciences, U.S. Department of Energy, under Contract DE-AC03-76SF-00098.

[2] The Spallation Neutron Source (SNS) project is a partnership of six U.S. Department of Energy Laboratories: Argonne National Laboratory, Brookhaven National Laboratory, Thomas Jefferson National Accelerator Facility, Los Alamos National Laboratory, Lawrence Berkeley National Laboratory, and Oak Ridge National Laboratory. SNS is managed by UT-Battelle, LLC, under contract DE-AC05-00OR22725 for the U.S. Department of Energy.

INTRODUCTION

The emittance of a particle beam describes its suitability for focusing and transport, an important characteristic for focused charged particle beam applications. The emittance is the six-dimensional distribution of all position coordinates along the three configuration-space directions and their associated velocity coordinates, normally measured from the center of the particle distribution. It is common to reduce the emittance to three easy-to-display subsets by projecting it into the two-dimensional planes {x-x'}, {y-y'}, and {z-z'}, respectively [1].

The emittance allows for the prediction of particle losses when the beam is accelerated, transported, or delivered onto a target. High-energy accelerators often encounter productivity limitations because of the radiation and activation generated by lost high-energy particles. Such particles dwell preferentially at the edge of the beam and can be removed before they gain excessive energy using carefully placed limiting apertures. This type of problem is best addressed with the size and shape of the emittance area that a certain fraction of the beam occupies in relation to the acceptance area of limiting apertures.

The emittance area, however, fails to consider the particle density distribution, which is rarely uniform across the beam, but normally peaks at the center. Accordingly, it is normally an under-proportionally small fraction of the particles that is at risk of being lost when changing the area of limiting apertures. On the other hand, the root mean square (rms) emittance takes the entire particle distribution into account and describes the focusability and transportability of a particle beam with one single, well-defined number [1-3]. Based on measured particle flux elements c(x,x') passing through a relative position coordinate x with a relative velocity component x', the rms emittance is defined as

$$\varepsilon = \sqrt{\langle x'^2 \rangle \langle x^2 \rangle - \langle xx' \rangle^2}$$

(1)

with $\langle x^2 \rangle = \dfrac{\sum_{all} x^2 c(x,x')}{\sum_{all} c(x,x')}$, $\langle x'^2 \rangle = \dfrac{\sum_{all} x'^2 c(x,x')}{\sum_{all} c(x,x')}$, and $\langle xx' \rangle = \dfrac{\sum_{all} xx' c(x,x')}{\sum_{all} c(x,x')}$

All terms above are normally evaluated after renormalizing x and x' so that the first moment becomes zero, namely,

$$\langle x \rangle = \frac{\sum_{all} x c(x,x')}{\sum_{all} c(x,x')} = 0 \;, \quad \text{and} \quad \langle x' \rangle = \frac{\sum_{all} x' c(x,x')}{\sum_{all} c(x,x')} = 0 \;.$$

(2)

The renormalization minimizes the calculated rms emittance, which corresponds to the normal process of maximizing transmission when steering a particle beam through a limiting aperture. The orientation and aspect ratio of the rms emittance ellipse are described by the Twiss parameters, namely,

$$\alpha = -\frac{\langle xx' \rangle}{\varepsilon}, \quad \beta = -\frac{\langle x^2 \rangle}{\varepsilon}, \quad \text{and} \quad \gamma = -\frac{\langle x'^2 \rangle}{\varepsilon}. \tag{3}$$

Many devices have been developed to measure the particle flux elements $c(x,x')$, such as pepper-pots, slit and collector scanners, double scanning slits, and electrical sweep scanners to name a few. All of the emittance measurement devices include a limiting entrance slit or aperture, which defines the initial position x of a particle flux element $c(x)$. The particles that pass through the entrance device are intercepted farther downstream by a second device, which measures the position distribution to determine the distribution of the corresponding velocity component x' of the particles with initial position coordinate x. Emittance measurement hardware and associated problems and limitations have frequently been discussed in the literature [1,3-9]. All such devices have a high level of complexity and therefore require a thorough checkout and calibration, preferably with a scanning pencil beam as well as no beam, to ensure reliable and accurate emittance data.

On the data reduction side, thorough discussions seem to be limited to theoretical distributions and to the "Gaussian" analysis, for which the emittance is plotted versus the logarithm of 1 minus the enclosed beam fraction [1,5,10]. The Gaussian analysis gives an excellent assessment of the distribution of the beam core, roughly the inner 90% of the beam. However, it is important to include the entire beam for a proper evaluation of the rms emittance because the outer 10% contribute significantly because of the relatively high values of their position and velocity coordinates. But these values are even higher for any net background and/or artifacts measured far from the beam, which therefore need to be removed from the data or excluded from the analysis to avoid their potentially large contribution to the rms emittance estimate. This separation is intrinsically difficult and normally requires delicate judgments, which are likely to produce biased estimates with errors that are difficult to quantify. This paper focuses on the task of extracting the most accurate rms emittance estimate from emittance data. This is best accomplished with the here-introduced SCUBEEx method, a self-consistent, unbiased elliptical exclusion analysis that combines traditional rms emittance analysis methods with simple statistical methods [11,12]. The SCUBEEx estimates are compared with estimates obtained with traditional methods.

EMITTANCE MEASUREMENTS

For all practical purposes, emittance measurements are double slit experiments that measure the partial particle flux for broad ranges of positions x and corresponding velocity components x'. When the slits are in positions where they let pass a part of the beam core, the measured particle flux is a small fraction, roughly 1%, of the flux of the entire beam. This fraction drops to the 10^{-4} range when measuring the beam halo, the small particle flux found outside the beam core. Going further away from the beam core, the halo gradually fades away until the measured data consist of pure background.

For this paper, we dissect the background into two components: the noise, fast-varying randomlike variations with a zero average, and the bias, the mean value, which is constant, at least locally. The bias can be zero and could vary slowly with the values of the position and velocity coordinates, although the analyses in this study are restricted to uniform biases.

Because of the gradual transition and because of the inherent noise, it is impossible to clearly distinguish between background data and data containing halo. Restricting the analysis to the core of the beam could circumvent the problem, but again would fail to yield meaningful estimates of the rms emittance.

Increasing the signal-to-noise ratio of the data can reduce the problem; for example, increasing the width of both slits increases the signal but decreases the resolution and so increases the systematic error of the rms emittance estimate [1,5]. Reducing the noise is a challenge, especially when the particle flux is determined by measuring the minute electrical currents, which are easily affected by electronic noise and biases. A high-quality current amplifier is essential [9] but still requires careful stability checks and careful zeroing. Equally important is an effective electromagnetic isolation of the amplifier and the current probe, which should include coaxial or tri-axial cables preferably with a low tribo-electrical dielectric. In addition, one should measure the distribution of the background in absence of beam. Subtracting the measured no-beam background data from the emittance data can eliminate the bias but increases the noise.

Current amplifiers commonly have a polarity switch to produce a positive output independent of the polarity of the input current. For this paper, we define positive as having the same polarity as the current observed when measuring the beam core, while negative refers to the opposite polarity.

In the absence of an actual current and a net bias, the current amplifier outputs background noise, small positive and negative data in a quasi-random sequence. As defined, the background noise data have a zero average and therefore do not contribute to the rms emittance because, on average, their contributions cancel each other out when weighted with the measured current $c(x,x')$. A positive bias, however, will cause an overestimation of the rms emittance because of the dominance of positive data, whereas a negative bias causes an underestimation of the rms emittance. Even a small bias can drastically affect the rms emittance estimate because the emittance data are normally dominated by background data and because the small bias current values are multiplied with a large range of x and x', up to the highest values included in the emittance measurement. Accordingly, an unbiased rms emittance estimate requires bias-free data. We will show that this can be accomplished simply by subtracting a self-consistently determined bias from the emittance data.

This paper focuses on emittance measurements obtained with two slits followed by a single current probe connected to a single current amplifier; therefore, all data of a set can be expected to have the same bias, if any. Thorough examinations frequently reveal a small net bias even for data obtained with a perfectly zeroed amplifier, as well as for data corrected with measured background data. Electronic drifts of the amplifier, an imperfect amplifier zeroing procedure, or background biases that differ between beam and no beam, can be the cause. Such biases are normally a very small fraction of the noise amplitudes and therefore often go unnoticed.

Figure 1 shows actual emittance data as a density plot versus position x and corresponding velocity component x'. This distribution was measured with a Penning type H- source operated at −35 kV on the ISIS ion source development rig. It is the horizontal complement to the vertical emittance data shown elsewhere in this volume [13]. The data describe an expanding beam without significant aberrations as seen in the tilted series of concentric ellipse-like areas of increasing darkness.

The figure shows both tails to be missing because the scan from −28 to 28 mm does not cover the full size of the beam. Tight scanning limits are often selected in a well-intended effort to save time, to optimize resolution, and/or to reduce analysis problems. Such incomplete emittance data, however, lead to underestimating the rms emittance and therefore are of questionable value. Nevertheless, such incomplete data sets are suited to demonstrating that reliable rms emittance estimates can be obtained, while being fully aware that the true rms emittance of this beam is significantly higher.

Calculating the rms emittance from all shown data yields 228 mm·mrad, which is the product of the two half-axes of the rms emittance ellipse, as defined in (1) and (3). In agreement with this definition, we use mm·mrad as unit rather than the often-used unit of π·mm·mrad, where the sometimes confusing π is intended to be a superficial reminder that the value has to be multiplied with π when comparing it with an emittance area.

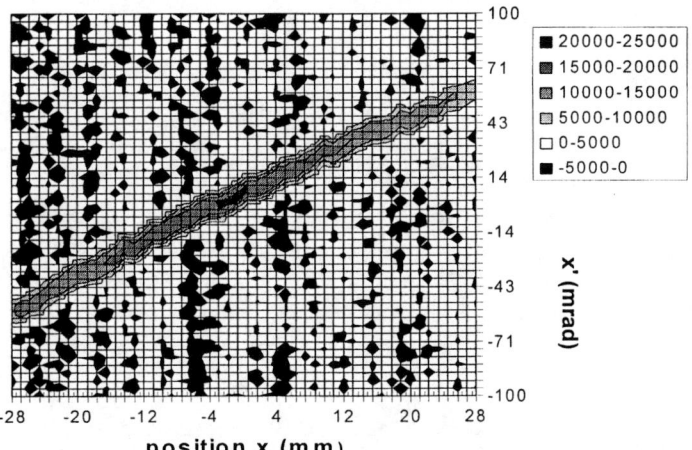

Figure 1. Emittance data from the ISIS development rig [13]. More than 85% are actual background data composed of small positive (white) and negative (black) current data. The SCUBEEx analysis uses this background to obtain unbiased, self-consistent estimates for the rms emittance and its uncertainty. The ends of the measured particle distribution are clipped causing the rms emittance to be underestimated.

HISTOGRAM ANALYSIS

Figure 1 shows the high-current emittance data surrounded by a narrow zone of small, exclusively positive data indicated in white. Only further away from the area with higher currents can one observe negative numbers as well, indicated in black. The distribution and pattern of the data far away from the beam core appear to be compatible with a field of quasi-randomly positive and negative small numbers, as one would expect from a background with noise but without a significant bias.

A better assessment of the background data can be obtained with a histogram when plotting the number of data that fall within a small current range versus the mean current in this range. Because the background data cover only a narrow current range, they form a peak close to zero, while the data containing actual particle beam are distributed over the entire range and become visible only on the high value side where they taper off towards the highest measured current of 20,888 arbitrary units.

In emittance measurements, the currents are commonly measured in arbitrary units because any determined calibration factor cancels out when evaluating the emittance. Accordingly, we give all current values as plain numbers without any unit. Only occasionally is a current compared with the highest measured current, which then is clearly indicated by the percentage unit.

The ISIS data in Figure 2 start at –300 and have a mode close to 50. Because this background noise is most likely symmetric or almost symmetric, it suggests that the background extends to roughly 400 with a mean about 50. This bias value is very small, corresponding to about 15% of the highest noise amplitude, or about 0.2% of the highest measured current.

Although the histogram analysis gives a good assessment of the dominating background, it cannot isolate the distribution of the small current data measured in the beam halo because it is partly buried under the background peak.

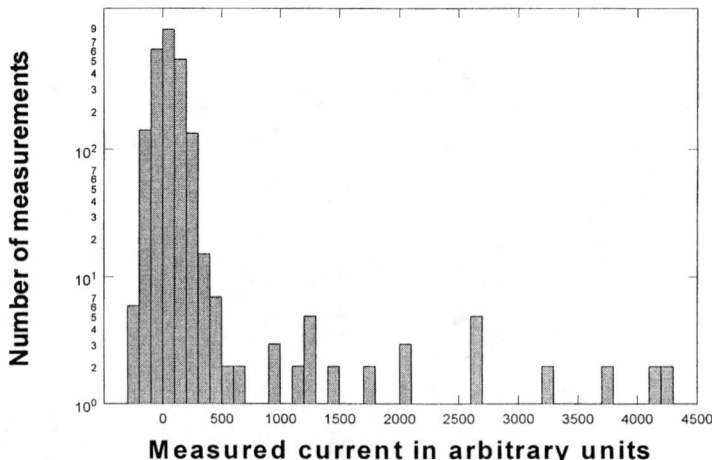

Figure 2. The histogram analysis of the ISIS emittance data shows the background data to be distributed near zero. The mode not being at zero indicates a small bias of about 50, or 0.2% of the highest measured current.

THRESHOLD ANALYSIS

The histogram suggests that rms emittance contributions from all background can be eliminated if the threshold for the data is set at roughly 400, meaning that all measured currents $c(x,x')$ below 400 are either set to zero or excluded from all summations. To standardize this method, we quote all thresholds as a fraction of the highest measured current, clearly indicated by the percentage unit. Accordingly, the 400 correspond to a 2% threshold, which reduces the rms emittance estimate to 64.5 mm·mrad.

The rms emittance estimate, however, depends on the selected threshold value. Figure 3 shows the rms emittance estimate for the ISIS data as a function of the selected threshold. Accordingly, the estimate grows rapidly when the threshold is lowered below 2% until it reaches a maximum of 270 mm·mrad, with the threshold at zero where all negative values are ignored. Gradually reinstating the negative values by lowering the threshold into the negative range lowers the rms emittance estimate. It reaches the previously established 228 mm·mrad when selecting a threshold below −1.3%, which includes all data. The rms emittance peak centered on zero is obviously the result of the positive background data not being compensated by the negative background data. The change of slope observed at 2% is sometimes used as threshold selection criterion, although it lacks a rigorous justification.

The threshold analysis and the histogram analysis sort the data according to the measured current. Because the background normally accounts for most of the data in the low current range, it dominates the results of these analyses. These dominating background features partly bury contributions from the few small, actual current measurements from the beam halo.

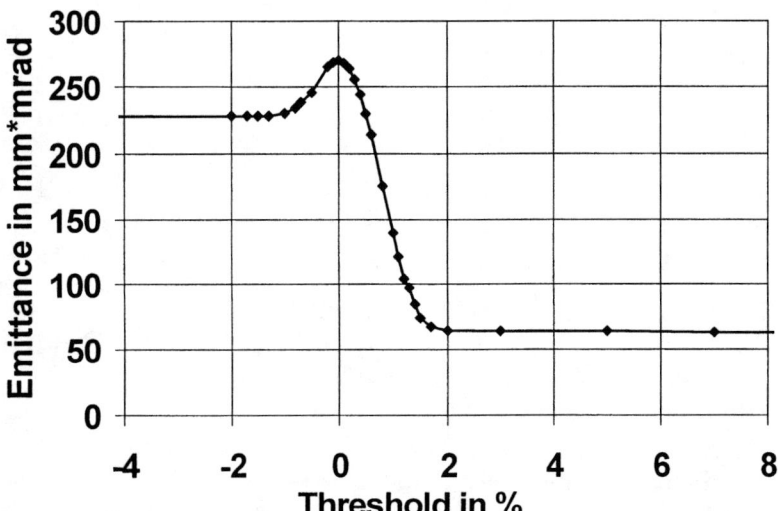

Figure 3. Threshold analysis of the ISIS emittance data shows the rms emittance estimate as a function of the threshold value expressed in a percentage of the highest measured current. The peak centered on zero is an artifact produced by excluding the negative numbers from the analysis.

ELLIPTICAL EXCLUSION ANALYSIS

The exclusion analysis ignores data outside a certain boundary (or sets them to zero) in order to exclude all data that are assumed not to be a part of the particle beam and that are called "unphysical" by some people. This method excludes predominantly background, while giving more weight to beam-halo data because it considers their vicinity to the beam-core coordinates.

The exclusion boundary should tightly surround all data that contain actual current to maximize the exclusion of the background. Exclusion boundaries can be of various shapes, but only elliptical boundaries are used in this study because they are easy to describe and can be closely fitted to many actual phase-space distributions. The threshold for the ISIS data, for example, was held at 5% before calculating the Twiss parameters α and β (3) to determine the orientation and aspect ratio of the exclusion ellipses used for Figure 4. This figure presents different quantities as a function of the half-axis-product (HAP), the product of the two half-axes of the exclusion ellipse, whether or not some part of the elliptical area falls outside the measured data field.

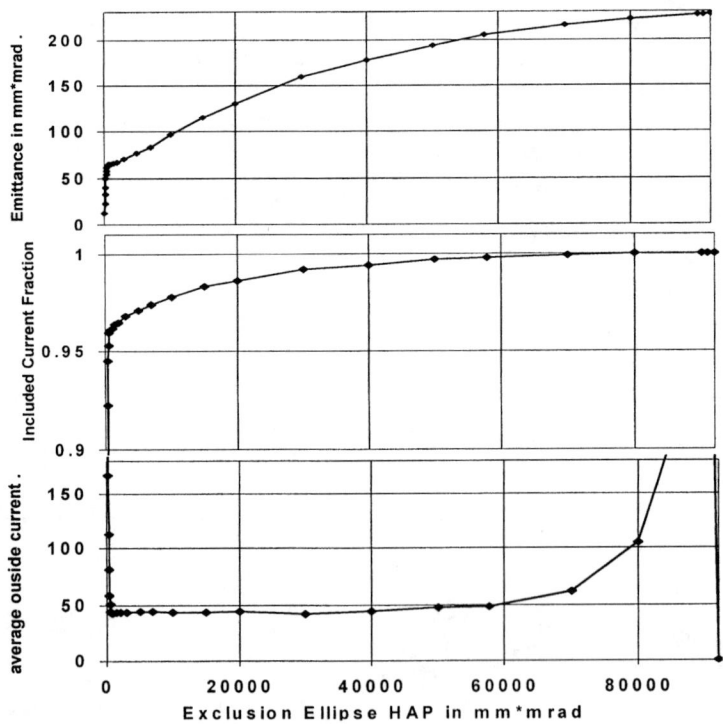

Figure 4. Exclusion analysis of the ISIS emittance data as a function of the HAP of the exclusion ellipse: (top) rms emittance estimate from the raw data within the ellipse; (center) fraction of total current within the ellipse; and (bottom) average current density outside the ellipse, which is consistent with a uniform bias current of roughly 45 in most areas. The orientation α and aspect ratio β of the exclusion ellipses were determined from all emittance data exceeding 5% of the highest measured current.

The top of Figure 4 shows the rms emittance estimate calculated from the data within the exclusion ellipse, which quickly reaches 65 mm·mrad with a HAP of 700 mm·mrad, but continues to grow up to the previously established 228 mm·mrad with a HAP of 92,000 mm·mrad when the entire data field is inside the exclusion ellipse. This analysis shows that a significant part of the rms emittance growth originates from data very far from the beam core with coordinates that are physically impossible to be populated by beam particles.

The center of Figure 4 shows the sum of the currents $c(x,x')$ measured inside the exclusion ellipse as a fraction of the sum of the currents measured for the entire data field. The percentage of current inside a 700 mm·mrad ellipse is 96%, while the 4% current outside that same ellipse accounts for a 3.5-fold increase in the rms emittance estimate.

The bottom of Figure 4 shows the current density (per data point) averaged over the excluded data outside the exclusion ellipse. The data show the average outside current to drop rapidly to 43 with HAP of 700 mm·mrad and to remain fairly constant up to 40,000 mm·mrad, where it starts to increase. As the exclusion ellipse is increased, the excluded area becomes rapidly smaller because it is restricted to two remaining corners, causing drastic changes. For this reason the following analyses are restricted to exclusion ellipses with a HAP of 20,000 mm·mrad or less.

Because the total of the actual particle beam current is limited, the current density of the beam halo has to decrease with increasing distance from the beam core, at least for large distances, and eventually reach zero. Accordingly, the plateau found on the bottom of Figure 4 for the average outside current measured outside 700 mm·mrad shows that these data cannot represent actual beam current but have to be background bias such as a dc-offset of the current amplifier.

Choosing an exclusion ellipse with a HAP of 700 mm·mrad results in an rms emittance estimate of 65 mm·mrad, but this estimate clearly ignores the fact that the analysis has identified a bias in the measured currents that results in a biased rms emittance estimate.

BIAS SUBTRACTION ANALYSIS

As discussed previously, any bias, as illustrated in Figure 5a, should be subtracted from the data to obtain an unbiased rms emittance estimate. Thresholding, illustrated in Figure 5b, does not eliminate the bias from the retained part of the distribution and tends to clip the tails of the distribution. Subtraction is the only method that can

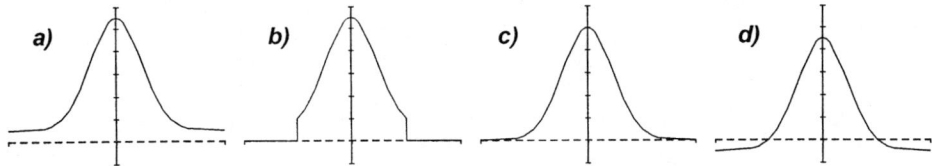

Figure 5. a) Distribution with a 10% background bias, b) after being thresholded at 20%, c) after subtracting 10% bias, and d) after subtracting 20% bias.

correctly eliminate a bias, as illustrated in Figure 5c.

However, when an excessive bias is subtracted, the method converts a significant fraction of the data to negative numbers, as illustrated in Figure 5d, which can lead to a gross underestimation of the rms emittance.

Figure 6 shows the rms emittance estimates of the bias-subtracted ISIS data as a function of the subtracted bias. Starting at the previously established 228 mm·mrad, the rms emittance estimate drops gradually when increasing the subtracted bias. The pace, however, accelerates with increasing bias subtraction, until a subtracted bias of 48.356 estimates the rms emittance to be zero, and imaginary beyond. This underlines the potential problem of the bias subtraction: according to Figures 4 and 6, the rms emittance estimates are imaginary if the bias is estimated from the data outside an ellipse with a HAP of 57,800 mm·mrad. Imaginary rms emittance estimates are clearly unphysical and contradict the assumption that the data farthest away from the actual beam yield the most reliable bias estimates.

Accordingly, reliable bias estimates are better obtained from data in the vicinity of the particle beam but clearly outside the beam halo, a boundary that can be established with the elliptical exclusion analysis. Exclusion ellipses with a HAP between 600 and 15,000 mm·mrad estimate the bias at 44.3 ± 1.5, which yields an rms emittance estimates in the range between 80 and 53 mm·mrad, not a very accurate estimate.

The rms emittance estimated from all bias-subtracted data is very sensitive to the subtracted bias because most data are background data (more than 85% of the ISIS data). A small error in the estimation of the subtracted bias leaves enough net bias to significantly change the rms emittance estimate. This problem can be reduced if the rms emittance is estimated from within a smaller area, which suggests combining the bias subtraction with an exclusion analysis.

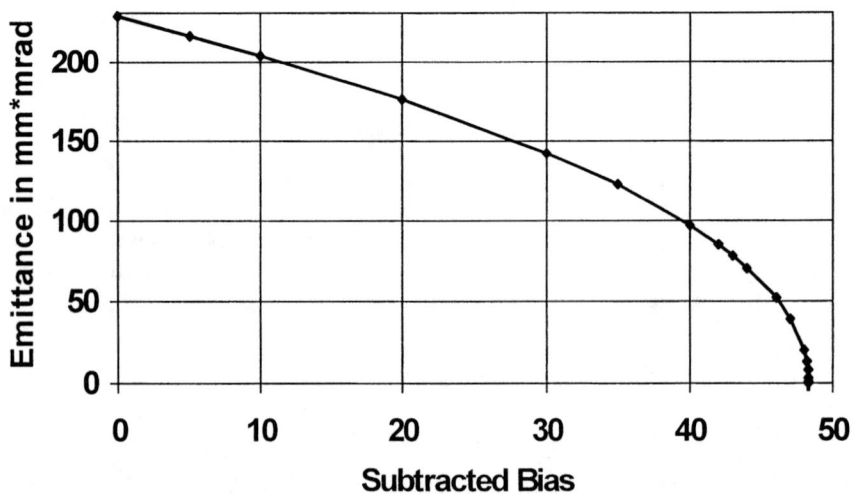

Figure 6. Bias subtraction analysis of the ISIS emittance data as a function of the subtracted bias. The dominance of background data makes the rms emittance estimate very sensitive to the subtracted bias, yielding imaginary estimates above 48.356.

SCUBEEx, THE SELF-CONSITENT, UNBIASED ELLIPTICAL EXCLUSION ANALYSIS

As seen previously, the elliptical exclusion analysis yields reasonable bias estimates from the average current measured outside the exclusion ellipse when the average-outside-current versus exclusion-ellipse-HAP plot exhibits a plateau after the initial drop. The bottom of Figure 4 and the top of Figure 7 represent the same average outside currents, but the latter display is expanded by a factor of 4 and features a finer evaluation. Almost all shown average outside current estimates above 550 mm·mrad are within a range of 44.3 ± 1.5, showing no significant trend, but only random-like variations.

The solid line in the bottom of Figure 7 shows the rms emittance estimated from the raw data within each exclusion ellipse, which initially grows rapidly as an increasing fraction of the actual particle beam is included in the growing ellipse. However, when the ellipse HAP reaches 550 mm·mrad, the rms emittance estimate starts to grow at a slower rate. The dashed line in the bottom of Figure 7 shows the rms emittance contribution from a uniform, constant current inside the exclusion ellipse, calculated from the corresponding average outside current shown in the top of the same figure.

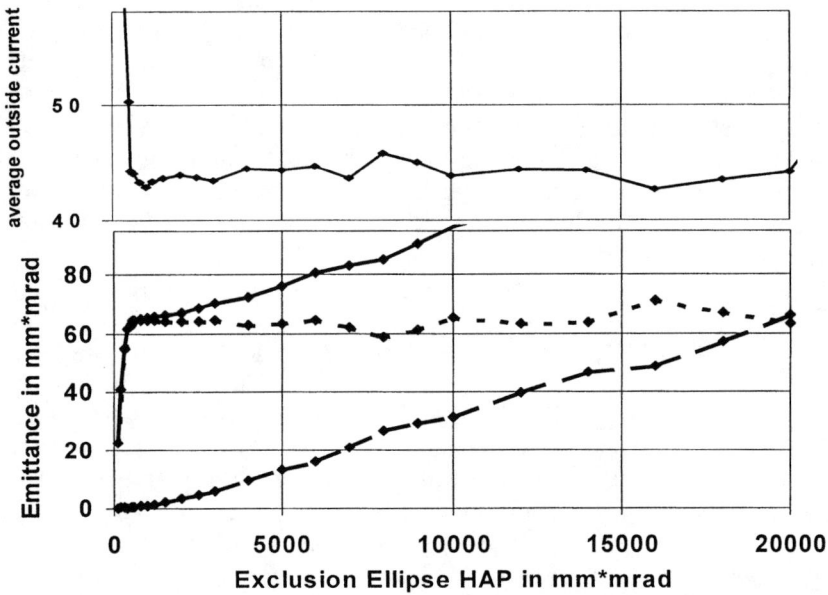

Figure 7. Unbiased exclusion analysis of the ISIS emittance data as a function of the HAP of the exclusion ellipse: top) average current density outside the ellipse (used as bias estimate) and bottom) rms emittance estimate calculated from the raw data (solid) and from the bias-subtracted data (dotted) inside the exclusion ellipse. The dashed line shows the rms emittance contribution from a uniform current equal to the estimated bias current. The rms emittance plateau reached with the unbiased data confirms reasonable bias subtraction and that all actual particle beam current was included. The ellipse parameters α and β were determined from all emittance data exceeding 5% of the highest measured current.

Starting at zero, this rms emittance contribution gains value approximately at the same rate as the rms emittance estimate based on the raw data for ellipses with a HAP in excess of 550 mm·mrad. This demonstrates that the growth of the rms emittance estimate for larger ellipses is caused by bias, which can be estimated from the average outside current, and has to be subtracted from the raw data.

Rather than subtracting the estimated bias from the data and calculating the rms emittance from all data, SCUBEEx calculates the rms emittance only from the bias-subtracted data within the exclusion ellipse, which reduces its sensitivity to errors in the estimated bias. For every evaluated ellipse, the SCUBEEx method first calculates the average current outside the evaluated ellipse. This average current is subtracted from all data to obtain unbiased data. The unbiased data within the ellipse are then used to calculate the unbiased rms emittance estimate. This unbiased rms emittance estimate is shown as a dotted line in the bottom of Figure 7. The estimate grows until it reaches a plateau when the exclusion ellipse exceeds a HAP of 550 mm·mrad. The fact that the rms emittance estimate reaches a plateau is a self-consistent confirmation that the bias has been properly subtracted and that the rms emittance estimate includes the contributions from all real current measurements, even if they were buried in the noise of individual current measurements.

To form a plateau, the unbiased rms emittance estimates have to be trendless over a broad range of exclusion ellipses, as seen in Figure 7. However, the figure also shows the estimate to feature quasi-random variations. As expected, many of the noiselike variations are anticorrelated with variations of the average outside current shown in the top of the same figure. The amplitudes of these variations grow with the exclusion ellipse because of increased sensitivity to the bias estimate, a problem already discussed for the bias subtraction analysis. Therefore, future evaluations of the ISIS emittance data will be restricted to a maximum HAP of 7000 mm·mrad. In the HAP range between 450 and 7000 mm·mrad, all unbiased rms emittace estimates are within the range of 63.4 ± 1.2, an uncertainty of roughly 2%.

ROBUSTNESS OF THE SCUBEEx ANALYSIS

The preceding analysis is based on varying the area of the exclusion ellipse while keeping the orientation and aspect ratio fixed. These two Twiss parameters were calculated from the data remaining after being thresholded at 5%, a somewhat arbitrary choice. Figure 8 shows these Twiss parameters α and β for the ISIS emittance data as a function of the selected threshold value. It shows α and β to undergo drastic changes between -1 and 2%, the range which is completely dominated by background according to the histogram analysis. Some people use the upper end of this dramatic change to locate the threshold that should exclude the entire background. The upper end at 2% seen in Figure 8 matches the change of slope observed in Figure 3, which was previously discussed as a possible selection criterion.

At higher thresholds, α and β change rather slowly, making the ellipse stretch and wobble just a little, as can be seen in Figure 9. This figure shows four ellipses with a

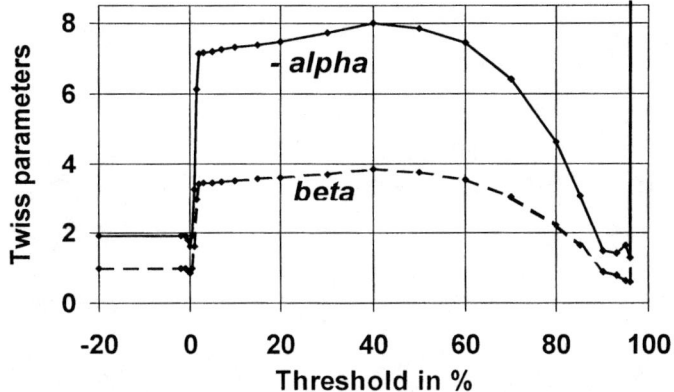

Figure 8. The threshold analysis of the ISIS emittance data shows the Twiss parameters α and β as a function of the threshold. Elimination of the background produces sharp changes between -1 and 2%.

HAP of 200 mm·mrad and with α and β thresholds of 5% (A), 40% (B), and 80% (C). Ellipse D is rather roundish because its parameters were determined with a threshold below −1.3% to include all data.

Figure 10 shows the SCUBEEx analysis repeated for exclusion boundaries scaled from the four ellipses shown in Figure 9. The dash-dotted lines are obtained by scaling ellipse A and therefore are the same data as in Figure 7 except that Figure 10 features a finer evaluation and is expanded by a factor of 3. The dashed lines, obtained by scaling the slightly elongated and twisted ellipse B, are practically on top of the dash-dotted lines, confirming the results obtained with ellipse A. A noticeable difference is observed from the dotted lines obtained by scaling ellipse C, which is less elongated. At 5000 and 5500 mm·mrad, the ellipse yields unusually high bias estimates that lead to unusually low rms emittance estimates. This appears to be the same background variation that can be observed in Figure 7 with ellipse A at a HAP of 8000 mm·mrad. If this depression is ignored, which is far from the beam, all other estimates obtained with ellipses A, B, and C are within the range of 63.8 ± 1.1. This range is in fair agreement with our previous estimate based exclusively on ellipse A, proving the robustness of the SCUBEEx method.

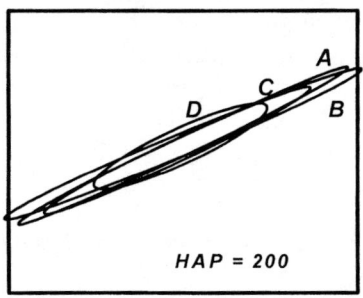

Figure 9. Ellipses with a HAP of 200 mm·mrad used for the robustness test of the unbiased elliptical exclusion analysis shown in Figure 10. Analyzing the ISIS emittance data that exceed 5% (A), 40% (B), or 80% (C) of the highest measured current, as well as analyzing all data (D), yielded the corresponding alpha and beta parameters.

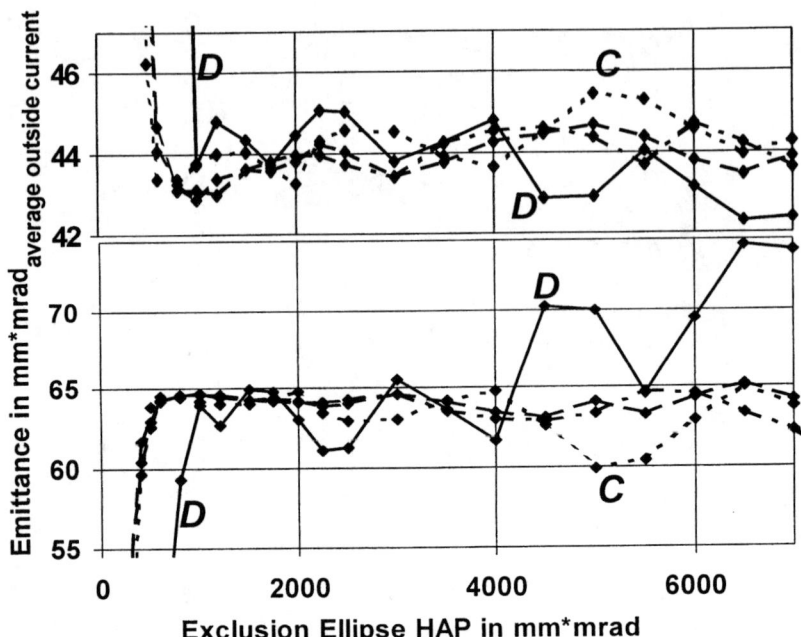

Figure 10. Unbiased exclusion analysis of the ISIS emittance data as a function of the HAP of different exclusion ellipses: top) average outside current, and bottom) rms emittance estimated from the bias-subtracted data inside the exclusion ellipses A (dash-dotted), B (dashed), C (dotted), and D (solid), shown in Figure 9. All ellipses give consistent estimates, but the variations increase in amplitude as the fit gets poorer with ellipses C and D.

The solid lines in Figure 10 are obtained with the roundish ellipse D calculated from all data. Because of its badly fitted aspect ratio, it takes a much larger ellipse to include all actual measured current, and therefore the average outside current and the unbiased rms emittance estimate reach their plateaus with a HAP of 1000 mm·mrad, almost twice the value needed for the other ellipses. For the same reason, the background variation discussed previously seems to appear at a HAP of 2400 mm·mrad, about two to three times smaller than with the other ellipses. These two effects shorten the plateau almost beyond recognition. If SCUBEEx were used exclusively with ellipse D, the rms emittance would probably be estimated with 63 ± 2, with a relatively low level of confidence. This example shows how the estimates benefit from selecting an exclusion boundary that tightly surrounds all emittance data that contain actual particle flux.

NEGATIVE NUMBER SUPPRESSION

According to our definition, all actual measured currents should be positive, and it is not surprising that negative current data are frequently questioned. A radical notion argues that all negative current data are unphysical and therefore ought to be eradicated by converting them to zero. When all negative data in emittance measurements are zeroed, their canceling benefit gets lost and the remaining positive results from the widespread background are likely to drastically inflate the rms emittance estimate. If this zeroing process is combined with an exclusion analysis, one is still likely to estimate a reasonable rms emittance because most zeroed numbers are excluded with the background and because the few, if any, zeroed numbers inside the exclusion boundary are unlikely to have a significant effect on the rms emittance estimate.

A strange situation evolves when the zeroing of negative numbers is preceded by a bias subtraction, as shown in Figure 11 for the ISIS emittance data. Starting without a bias subtraction at the previously established 270 mm·mrad, the rms emittance estimate decreases rather slowly when increasing the subtracted bias. When subtracting 50, the bias estimated with Figure 2, the rms emittance is estimated at 211 mm·mrad.

Only when the subtracted bias is increased to 300 is a more reasonable estimate of 65.1 mm·mrad obtained. Around this point, the curve in Figure 11 exhibits a change of slope, which could be used as a selection criterion for this method. For higher bias subtractions, the rms emittance estimate barely changes, for example, subtracting 500 yields 63 mm·mrad and subtracting 800 yields 62 mm·mrad. Even so the subtracted bias exceeds the actual bias by a huge factor, the rms emittances estimates never become imaginary because all negative numbers are suppressed. The absence of such warning bells and the insensitivity of the rms emittance estimate to the level of bias subtraction, as seen in Figure 11, gives this estimate an image of stability and accordingly a false image of credibility.

To obtain the reasonable rms emittance estimate of 65.1 mm·mrad, a bias of 300 has to be subtracted, which is very close to the upper limit of the noise data, 400, which was estimated in the histogram analysis. Such a high bias subtraction is necessary because practically all background data need to be eliminated in order to eliminate their unbalanced contributions to the rms emittance estimate.

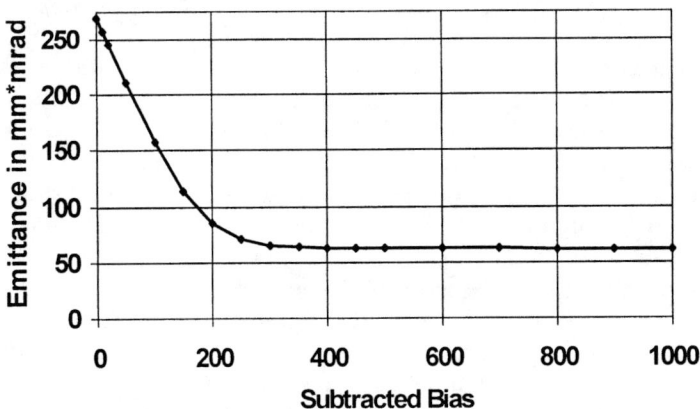

Figure 11. Bias subtraction followed by zero thresholding analysis of the ISIS emittance data as a function of the subtracted bias. Forfeiting the balancing effect of the negative current measurements, the method requires very large bias subtractions to obtain reasonable rms emittance estimates.

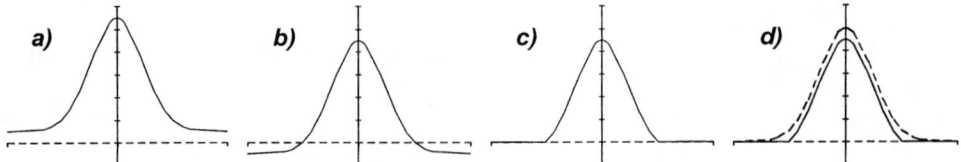

Figure 12. a) Distribution with 10% background bias, b) after subtracting 20% bias, c) after zeroing all negative numbers, and d) case c) again (solid line), in comparison with the original distribution after the proper 10% bias subtraction (dashed line).

Subtraction of an excessive bias followed by zeroing the negative numbers is illustrated in Figure 12. This figure shows how the method reduces all measured currents by the excessive bias, which is a minor problem for the high current values, but overproportionately reduces the small current values up to complete elimination. As shown in Figure 12d, this eliminates the tails of the shown distribution, and therefore, most of particle beam halo and a significant part of particle beam wings. A further increase of the subtracted bias is unlikely to yield a significant change in the rms emittance estimate because the data have practically been stripped to the core of the beam. This discussion, as well as Figure 12d, shows that this method is likely to underestimate the rms emittance up to a significant fraction.

Even though this method can give reasonable rms emittance estimates for the ISIS data, it cannot be trusted because it preferentially eliminates beam halo and wings. Negative emittance data should be a frequent occurrence when measuring with a well-zeroed current amplifier because in most positions there is no actual current and in many other positions the measured actual current is smaller than the typical noise. These negative numbers are a natural part of measuring zero or small currents in a physical world, which always features some noise, and therefore the negative numbers have to be included in any statistically sound rms emittance analysis.

ANALYSIS OF THE LBNL EMITTANCE DATA

The Lawrence Berkeley National Laboratory (LBNL) emittance data are very different from the ISIS data and are therefore a complementary test for any emittance analysis. LBNL developed a radio-frequency (rf)-driven, cesium-enhanced bucket source for SNS with a 7-mm extraction aperture, producing up to 50 mA of 65 kV H- [14]. The source is attached to a 10-cm-long low-energy beam-transport system (LEBT), which features two electrostatic lenses to tune the ion beam for optimal injection into the rf quadrupole (RFQ). Choosing a highly compact LEBT requires small-aperture, electrostatic lenses with inherent aberrations that cause the substantial curvature in the phase space distribution seen in Figure 13. This figure shows the same data that are shown in reference 14 as "vertical" data, with the normalized rms emittance estimated at 0.15 mm·mrad after "subtracting a 1.3% background." This corresponds to 13 mm·mrad unnormalized, the standard used in this study.

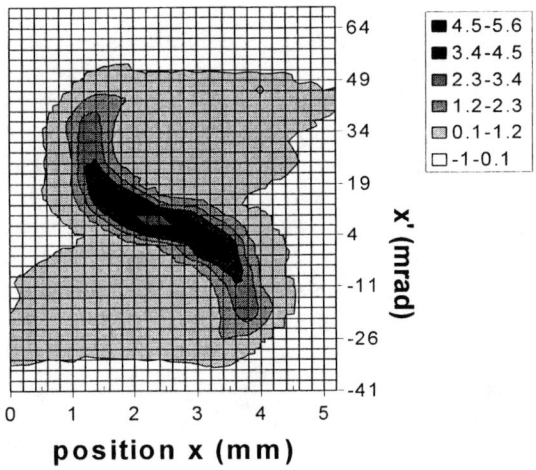

Figure 13. The measured LBNL phase space distribution obtained at 33 mA exhibits significant curvature because of large aberrations. Unfortunately the ends of the tails were not measured.

This emittance was measured behind an 8-mm LEBT exit aperture substituting for the actual RFQ entrance aperture and therefore closely represents the emittance of the beam injected into the RFQ. The emittance data are shown in a darkening gray scale starting with white at zero, except for the intermediate gray for the highest currents between 4.1 and the peak current of 6.85. The LBNL emittance data show the core of the beam to be converging, although the low-intensity wings have components anywhere between converging and diverging.

Figure 13 shows that the extreme parts of the wings are truncated. Accordingly, a small fraction of particle beam is missing, causing an underestimation of the emittance. This figure also shows that the actual particle beam fills almost the entire data field, and therefore rather little background is included. The raw data shown in Figure 13 yield an rms emittance estimate of 16.4 mm·mrad.

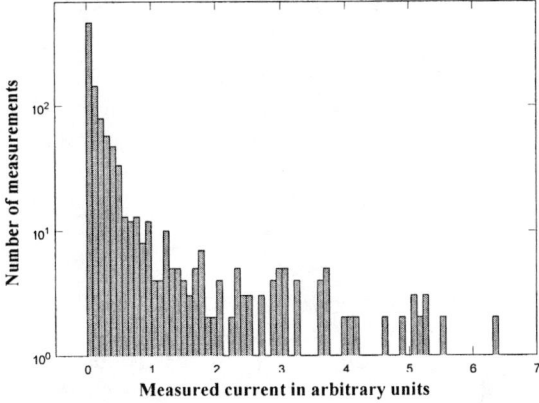

Figure 14. The histogram of the LBNL emittance data shows no negative values because the data acquisition program zeroed them.

The histogram shown in Figure 14 reveals the absence of any negative data because they are set to zero by the data acquisition program used with the LEBT emittance device at LBNL. A closer inspection of the raw data shows that only about 40% of the background data are exactly zero, which indicates the presence of a small, positive bias. Therefore, the rms emittance must be slightly smaller than the estimate calculated from the raw data. On the other hand, the bias has to be significantly smaller than the bias estimated from the average of the background data, and therefore SCUBEEx will underestimate the rms emittance.

The histogram demonstrates how difficult it is to separate the actual small currents from the background data. The sharp drop between bins 6 and 7 could indicate the background endpoint. The corresponding 8% threshold estimates the rms emittance at 8.33 mm·mrad. However, a closer inspection of the raw data shows that all background data are contained in the first bin. Thresholding the data at 1.3%, the upper limit of the first bin, estimates the rms emittance at 15.4 mm·mrad.

The threshold analysis for the Twiss parameters α and β is shown in the top of Figure 15. The α parameter changes significantly over the range between 0 and 15%. If the upper end of this range, 15%, is selected as a threshold to exclude the entire background as previously discussed, the rms emittance estimate drops to 6.71 mm·mrad.

The rms emittance estimate, however, depends strongly on the selected threshold, as seen in the bottom of Figure 15. Because all negative data were zeroed, the emittance estimate remains at the highest value of 16.4 mm·mrad for negative thresholds, having lost the compensating benefit of the negative data. This figure shows a change of slope at roughly 8%, which corresponds to an rms emittance estimate of 8.33 mm·mrad.

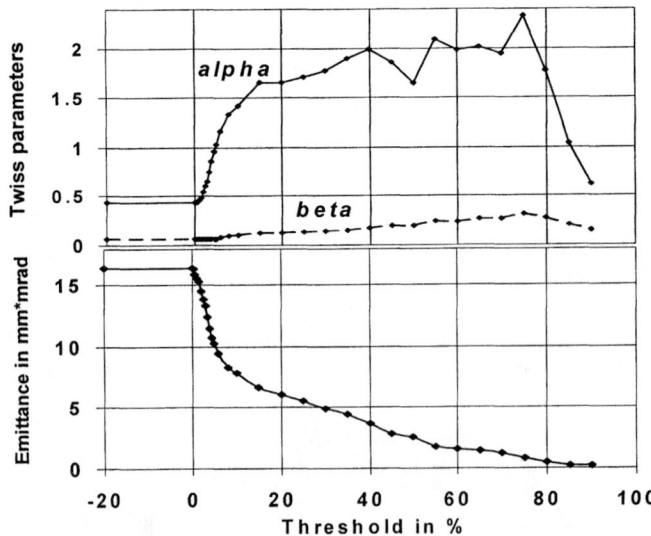

Figure 15. Threshold analysis of the LBNL emittance data shows top) the Twiss parameters alpha (solid line) and beta (dashed line) and bottom) the rms emittance estimate as a function of the threshold.

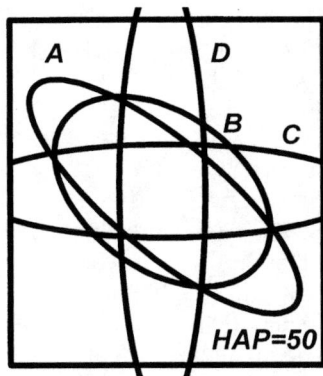

Figure 16. Ellipses with a HAP of 50 mm·mrad used for the robustness test of the unbiased elliptical exclusion analysis shown in Figure 17. Ellipses A and B are based on Twiss parameters evaluated from thresholded data, while C and D are designed to line up with the major axes.

The wide range of rms emittance estimates obtained with different methods of selecting a threshold asks for an unbiased estimate. Figure 16 shows four ellipses, all with HAP of 50 mm·mrad, which are used for the SCUBEEx analysis shown in Figure 17.

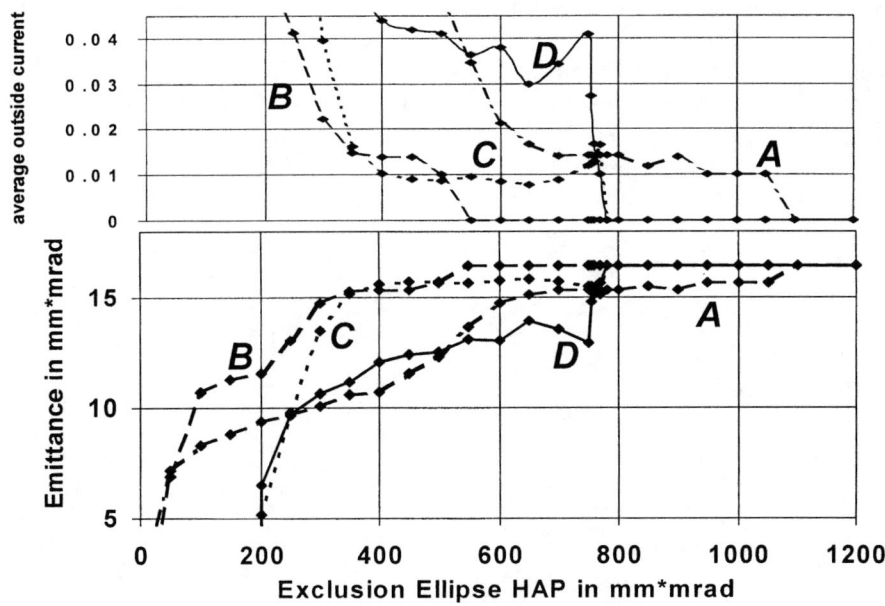

Figure 17. Unbiased exclusion analysis of the LBNL emittance data as a function of the HAP of different exclusion ellipses: top) average outside current, and bottom) rms emittance estimated from the bias-subtracted data inside the exclusion ellipses A (dash-dotted), B (dashed), C (dotted), and D (solid). The A, B, and C ellipses yield plateaus consistent with a bias of 0.011 ± 0.003 and an rms emittance of 15.5 ± 0.3. The plateaus obtained with ellipse D are caused by the clipped distribution tails and are therefore misleading.

The perfectly flat rms emittance plateaus at 16.4 mm·mrad accompanied by zero average outside current in Figure 17 are artifacts, which are found when the exclusion ellipses include all measured data. Of relevance are the (somewhat noisy) plateaus just to the left of the trivial plateaus.

Ellipse A is calculated with the Twiss parameters obtained from all data exceeding 10% of the highest measured current. Accordingly, although it fits the core of the beam well, it fits the distribution wings poorly and therefore takes a relatively large HAP of 700 mm·mrad before the dash-dotted lines reach the relevant plateaus. A slightly different background bias in different corners of the data field likely causes the small change of the plateau levels at 950 mm·mrad.

Ellipse B is calculated with the Twiss parameters obtained from all data and is therefore better suited to include all actual current data more rapidly with a minimum exclusion ellipse HAP of 350 mm·mrad. Accordingly, the corresponding dashed lines are very similar to the dash-dotted lines from ellipse A except that corresponding features are found at about half the HAP values. Unfortunately, ellipses A and B are able to isolate only the background in the upper right and lower left corners of the data field, which represents only a small fraction of all acquired background data.

Ellipse C was designed to simultaneously exclude the entire background on the top and bottom of the data field, and therefore the associated dotted lines feature the most pronounced plateaus.

Ellipse D was designed to demonstrate a potential problem: the truncated distribution wings result in plateaus, shown in solid lines, which could easily be mistaken to indicate the onset of background and thus lead to an obviously incorrect rms emittance estimate of 13 mm·mrad.

The more suitable ellipses A, B, and C produce plateaus all consistent with a bias of 0.011 ± 0.003 and an rms emittance of 15.5 ± 0.3 mm·mrad, an uncertainty of only 2%. This estimate differs significantly from all other estimates obtained with different thresholds selected according to different criteria.

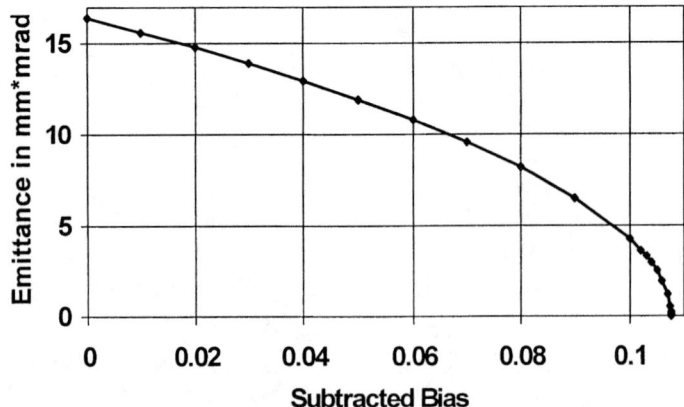

Figure 18. Bias subtraction analysis of the LBNL emittance data as a function of the subtracted bias.

Figure 18 shows the rms emittance estimate as a function of the subtracted bias. Using the SCUBEEx bias estimate, one obtains an rms emittance estimate of 15.55 ± 0.25 mm·mrad, in excellent agreement with the SCUBEEx estimate.

Subtracting a bias of 0.09 (1.3% of the highest measured current) estimates the rms emittance at 6.5 mm·mrad. To obtain an imaginary emittance estimate, one has to subtract a bias of 0.10763, which is 2.4 times larger that the highest value found in a closer inspection of the background data. The LBNL rms emittance estimates are relatively insensitive to the subtracted bias because the data occupy a rather small mm·mrad area.

Figure 19 shows the rms emittance estimate if zeroing of the negative numbers follows the bias subtraction. Using the 0.011 SCUBEEx bias estimate, one obtains an rms emittance estimate of 15.8 mm·mrad. Subtracting a bias of 0.09 (1.3% of the highest measured current) before clipping the negative numbers estimates the rms emittance at 13 mm·mrad, obviously the method of choice in reference 14.

The bias-subtraction-followed-by-negative-number-clipping estimate shows rather gradual changes. The most pronounced change of slope occurs between 0.2 and 0.4, which is about 4.5 to 9 times larger than the highest value found in a closer inspection of the background data. The corresponding rms emittance estimates are 10.2 and 7.6 mm·mrad, respectively. Clearly, this method yields rms emittance estimates, which scatter over a wide range. Most of these estimates severely underestimate the rms emittance, as predicted in the previous discussion.

Most methods yield very poor estimates of the LBNL rms emittance because they are unable to distinguish between the background and the small currents measured in the distribution wings and beam halo. Only SCUBEEx is able to isolate the LBNL background; therefore, only if its bias estimate is used in a bias subtraction analysis, or with the SCUBEEx method, can one obtain reliable estimates.

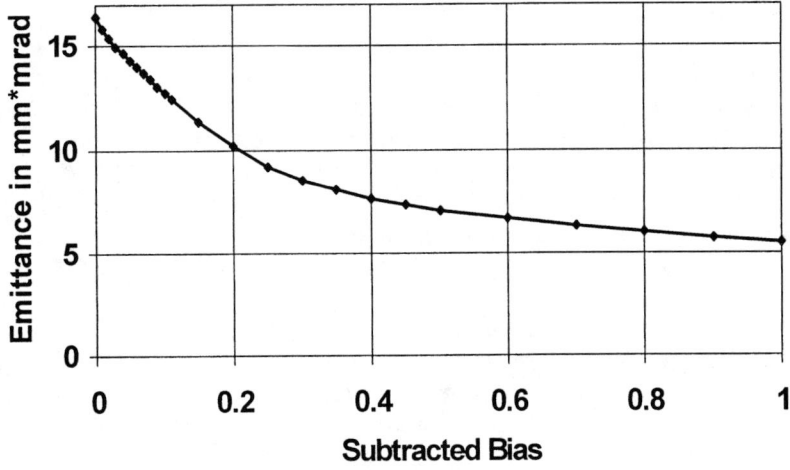

Figure 19. Bias subtraction followed by zero thresholding analysis of the LBNL emittance data as a function of the subtracted bias.

As discussed previously, the zeroing of negative numbers by the LBNL data acquisition system causes an overestimation of the SCUBEEx bias and therefore an underestimation of the rms emittance. On the other hand, the presence of the discussed small, positive bias leads to an overestimation of the rms emittance when evaluated from the raw data. These two facts can be reconciled by estimating the LBNL rms emittance at 16 ± 0.5 mm·mrad.

DISCUSSION

The threshold and histogram analyses sort the data according to the measured currents and therefore are dominated by the background data. It is practically impossible to clearly identify features that could be caused by the much smaller fraction of data related to the small measured halo currents. Systematic threshold analyses reveal features that in some cases can be used to obtain reasonable rms emittance estimates, but it remains unclear how well such thresholds relate to the different components of the original distribution.

The exclusion analysis uses inherent information on the x and x' coordinates of the actual distribution to separate the actual current data from the background data. It is therefore better suited to analyze the background and determine its bias, if any. Using a carefully selected exclusion boundary, one can identify small measured halo currents by comparing the average measured currents with the average measured background. However, the exclusion analysis by itself yields rms emittance estimates that depend on the actual background bias.

The SCUBEEx method estimates the bias from the data outside a carefully selected exclusion boundary before estimating the rms emittance from the bias-subtracted data within the boundary. When the exclusion area is increased, the bias as well as the rms emittance estimates normally reach plateaus, which is the self-consistent confirmation that all actual current data are inside the exclusion boundary. The uncertainty of the estimates can be estimated from the randomlike, trendless variations within the plateaus, which are caused by variations of the local average of the background data. When the exclusion area is further increased, the bias is estimated from a decreasing number of background data, and therefore the amplitudes of the observed variations increase. The increasing variations of the bias estimate cause increased variations of the rms emittance estimate. Accordingly, the evaluations of the plateaus ought to be limited to the vicinity of the beam to obtain accurate bias and rms emittance estimates.

The SCUBEEx method is somewhat robust with respect to the shape and orientation of the exclusion boundary. However, for the same aforementioned reason, the most accurate estimates are obtained when selecting the smallest possible exclusion area, which contains all data that feature actual particle beam current clearly exceeding the highest background data. For rather straight distributions, such an ellipse can be constructed using the Twiss parameters of the distribution calculated from the data thresholded above all background. Strongly curved distributions have been successfully subjected to this method as well, and the results show a surprisingly strong robustness. Improved credibility and accuracy can be expected when fitting the smallest exclusion ellipse that contains all data that feature actual particle beam current clearly exceeding the highest background data. A fitting procedure to accomplish this is given elsewhere [15].

Biases that vary slowly with the x and/or x' coordinates and biases that in an isolated area significantly deviated from the average bias can lead to sloped plateaus. Such challenges can be met by estimating the bias from the average background data inside an elliptical ring surrounding the exclusion ellipse. The ring thickness needs to be optimized and actually could serve as a standard robustness test.

SCUBEEx can be used for emittance data with and without a background bias. The ISIS and LBNL data were both measured with a single current amplifier, and therefore each set was subject to only one average bias value. Interpreting the estimated bias as average bias of many amplifiers, the method has been successfully used to analyze data obtained with a multistrip detector [16]. More reliable estimates could be obtained by developing an expanded SCUBEEx method that allows for different biases for different subsets of the emittance data.

Obtaining unbiased rms emittance estimates and their uncertainty from complete data sets (including the full distribution tails) allows for fair comparisons between different ion sources [16]. The incomplete sets still allow for comparing different analysis methods. Table 1 lists the estimates obtained with easily quantifiable methods discussed previously for the ISIS and LBNL data.

The ISIS data were measured with an average bias of approximately 464. To make the data more similar to other analyzed data, the bias was reduced by subtracting 420 from each measured value. Although this drastically reduced the rms emittance estimated from the raw data, it did not significantly change any other estimate listed in Table 1. Consequently, both data sets have a bias of approximately 0.2% of the highest measured current, respectively. The ISIS data, however, were taken over a 19.4 times larger emittance area, containing 90% background versus the 40% background of the smaller emittance area taken at LBNL. For these reasons, the raw

rms emittance estimates for:	ISIS data mm·mrad	LBNL data mm·mrad
Method:		
SCUBEEx	63.8 ± 1.1	15.5 ± 0.3
Raw data	228	16.4
Bias subtraction	66.5 ± 13.5 (44.3 ± 1.5)	15.5 ± 0.3 (0.011 ± 0.003)
Bias subtraction, negative numbers suppressed	65.1 (300)	8.5 (0.3)
Threshold, histogram	64.5 (2%)	8.3 (8%)
Threshold, change of slope	64.5 (2%)	8.3 (8%)
Threshold, change of α	64.5 (2%)	6.7 (15%)

Table 1. The rms emittance estimates for the ISIS and LBNL data in mm·mrad are listed for easy quantifiable analysis methods, mostly threshold analysis. The subtracted biases and the applied threshold is shown in parenthesis.

data overestimate the LBNL rms emittance by only 6% compared with 260% for the ISIS data.

Bias subtraction analyses yield smooth, featureless curves. But when combined with an extensive elliptical exclusion analysis to obtain self-consistent bias estimates, the bias subtraction gives credible rms emittance estimates for both cases. Small uncertainty exists for the LBNL data, but a large uncertainty for the ISIS data is the result of its much larger measured emittance area.

The ill-conceived suppression of all negative numbers preceded by bias subtraction yields smooth curves with at least one change of slope. Using the most prominent change of slope as criterion, the rms emittance estimate is credible for the ISIS data but underestimates the LBNL rms emittance by 50% because numerous low current data are mistaken as background data.

All threshold analyses give amazingly consistent and good estimates for the ISIS rms emittance, but underestimate the LBNL rms emittance by up to 57% for the same aforementioned reason.

Except for the straight evaluation from the raw data, all methods give reasonable estimates for the rms emittance of the ISIS beam. Except for the straight evaluation from the raw data, a careful bias subtraction, and SCUBEEx, all methods fail to give reasonable estimates for the rms emittance of the LBNL beam. SCUBEEx is the only method known to the authors that can unambiguously and accurately estimate the rms emittance together with credible estimates for the corresponding uncertainty for either data set. The uncertainties found to be in the few percent level for these data are much smaller than originally anticipated. Accordingly, one needs to consider additional uncertainties from other potential error sources [1].

CONCLUSIONS

Accurate and reliable rms emittance estimates require separation of the background from the real measured currents, which have to include the small currents measured in the beam halo. The common thresholding- and exclusion-analysis methods typically use a change of slope to select the optimal separation parameter. Because there is no rigorous justification, the exact selection of the separation parameter is normally based on intuition and experience, leaving some latitude for interpretation. Despite the inherent ambiguity, these methods appear to yield reasonable rms emittance estimates for well-defined data, but they appear to significantly underestimate the rms emittance of data containing numerous low current data and/or featuring a very noisy background.

The SCUBEEx method combines beneficial features of elliptical exclusion analysis and bias subtraction with simple statistical methods to obtain accurate rms emittance estimates from any measured emittance data. This method is the only one known to the authors that rigorously attempts to get unbiased estimates while simultaneously assessing their uncertainty caused mostly by variations in the background data. These rms emittance estimates include contributions from all statistically identifiable measured particle flux while excluding all contributions from statistical identifiable background. Of course, single parameter characterizations of phase-space distributions might not be sufficient for conducting detailed beam transport and acceleration studies. However, the unambiguous statistical identification of small particle currents and of pure background will without a doubt lead to more accurate simulations. And clearly, the improved accuracy of rms emittance estimates enables more reliable comparisons between different particle-beam-producing systems.

ACKNOWLEDGMENTS

This detailed study of rms emittance analysis has benefited significantly from discussions with our colleagues Alexander Aleksandrov, Reinard Becker, Stuart D. Henderson, Norbert Holtkamp, Miguel Olivo, Dave Olson, Ben A. Prichard, Jr., Paul Schmor, Ken Reece, Joseph D. Sherman, John W. Staples, James E. Stovall, Eugene P. Tanke, and Marion M. White. Special thanks go to Rudy Damm and Paul Gibson for providing the relief from other duties needed to make this work possible.

REFERENCES

1. LeJeune, C., and Aubert, J., in *Applied Charged Particle Optics*, edited by A. Septier, Academic Press, New York, 1980, pp. 159-259.
2. Keller, R., in *The Physics and Technology of Ion Sources*, edited by I. G. Brown, John Wiley and Sons, New York, 1989, pp. 23-52.
3. Zhang, H., in *Ion Sources,* Springer, Berlin, 1999, pp. 58-65 & 432-446.
4. Strehl, P., in *Handbook of Ion Sources*, edited by B. Wolf, CRC Press, New York, 1995, pp. 385-452.
5. Allison, P., in *Production and Neutralization of Negative Ions and Beams*, edited by J. Alessi, American Institute of Physics, New York, 1987, pp. 465-481.
6. Alessi, J., in *Production and Neutralization of Negative Ions and Beams*, edited by A. Hershcovitch, American Institute of Physics, New York, 1990, pp. 526-533.
7. Yuan, D., Jayamanna, K., Kuo, T., McDonald, M., and Schmor, P., *Rev. Sci. Instrum.* **67**, 1275-1276 (1996).
8. Hamabe, M., Kuroda, T., Sasao, M., Nishiura, M., Wada, and M., Guharay, S.K., *Rev. Sci. Instrum.* **71**, 1104-1106 (2000).
9. Dietrich, J., Mohos, I., Häuser, J., Riehl, G., and Weander, F., *Proceedings of the 8th European Particle Accelerator Conference,* 2002, pp. 1864-1866.
10. Holmes, A. J. T., in *The Physics and Technology of Ion Sources*, edited by Brown, John Wiley and Sons, New York, 1989, pp. 53-106.
11. Eadie, W.T., Drijard, D., James, F.E., Roos, M., and Sadoulet, B., *Statistical methods in Experimental Physics,* North-Holland, Amsterdam, 1971.
12. Rabinovich, S., *Measurement Errors, Theory and Practice*, American Institute of Physics, New York, 1995.
13. Thomason, J. W. G, in these proceedings.
14. Keller, R., Thomae, R., Stockli, M., and Welton, R., in these proceedings.
15. Keller, R., Sherman, J.D., and Allison, P., *IEEE Trans. Nucl. Sci.* **32**, 2579-2581 (1985).
16. Welton, R. F., Stockli, M.P., Keller, R., Thomae, R.W., Thomason, J., Sherman, J., and Alessi, J., in these proceedings.

Emittance Characteristics of High-Brightness H⁻ Ion Sources[1][2]

R.F. Welton[a] and M.P. Stockli[a], R. Keller[b] and R. W. Thomae[b],
J. Thomason[c], J. Sherman[d] and J. Alessi[e]

[a]*Spallation Neutron Source, Oak Ridge National Laboratory, Oak Ridge, TN, 37830-6473*

[b]*Lawrence Berkeley National Laboratory, Berkeley, CA 94720*

[c]*ISIS, Rutherford Appleton Laboratory, Chilton, Didcot, Oxon, OX11 0QX, UK*

[d]*Los Alamos National Laboratory, Los Alamos, NM 87545 USA*

[e]*Brookhaven National Laboratory, Collider Accelerator Dept. Bldg. 930, Upton, NY 11973-5000*

Abstract. A survey of emittance characteristics from high-brightness, H⁻ ion sources has been undertaken. Representative examples of each important type of H⁻ source for accelerator application are investigated: A magnetron surface plasma source (BNL); a multi-cusp-surface-conversion source (LANL); a Penning source (RAL-ISIS) and a multi-cusp-volume source (LBNL). Presently, comparisons between published emittance values from different ion sources are difficult largely because of different definitions used in reported emittances and the use of different data reduction techniques in analyzing data. Although seldom discussed in the literature, rms-emittance values often depend strongly on the method employed to separate real beam from background. In this work, the problem of data reduction along with software developed for emittance analysis is discussed. Raw emittance data, obtained from the above laboratories, is analyzed using a single technique and normalized rms and 90% area-emittance values are determined along with characteristic emittance versus beam fraction curves.

---

[1] This work is supported by the Office of Science, Office of Basic Energy Sciences, U.S. Department of Energy, under Contract DE-AC03-76SF-00098.

[2] The Spallation Neutron Source (SNS) project is a partnership of six U.S. Department of Energy Laboratories: Argonne National Laboratory, Brookhaven National Laboratory, Thomas Jefferson National Accelerator Facility, Los Alamos National Laboratory, Lawrence Berkeley National Laboratory, and Oak Ridge National Laboratory. SNS is managed by UT-Battelle, LLC, under contract DE-AC05-00OR22725 for the U.S. Department of Energy.

INTRODUCTION

The production of reliable, high-brightness beams of H⁻ has made possible a variety of accelerator injection, extraction and acceleration techniques, which are currently employed in accelerator facilities. For example, beams of H⁻ ions are employed along with beams of H⁺ ions for simultaneous acceleration in linacs; H⁻ beams can be efficiently extracted from cyclotrons with very low emittance growth using simple stripping techniques; and stripped H⁻ ions are used as the preferred method of injecting protons into high-current storage rings and high-energy synchrotrons[1]. Stripped H⁻ injection into an accumulator ring or synchrotron allows for multi-turn injection, through the stripping process, allowing increasing space charge density while preserving the total emittance of the beam to the highest degree possible[2].

Future facilities, such as spallation neutron sources, neutrino and muon factories and facilities devoted to the study of high energy collisions, will continue to utilize H⁻ production technology with increasing demands on ion source performance including: increased H⁻ beam current and duty factor, increased source reliability/availability and reduced emittance[2]. As beam energies and currents increase more, importance is placed on utilizing ion sources with the lowest possible intrinsic emittance, maximizing deliverable beam while minimizing activation of downstream accelerator components resulting in increased machine productivity.

This report presents a summary of emittance values and characteristic emittance-intensity curves from a representative sample of the current generation of state-of-the-art, high-brightness, H⁻ ion sources currently used or soon to be used as workhorses in accelerator facilities. Raw emittance data was obtained from each participating laboratory and includes most major types of high-brightness, H⁻ sources: The ISIS-RAL Penning source; the BNL magnetron source; the LBNL-SNS RF-driven-multi-cusp source and the LANSCE multi-cusp-converter source (LANL). In each case, analysis of the raw emittance data was performed using self-consistent, unbiased elliptical exclusion analysis (SCUBEEx)[3] applied by a computer program to best account for background and noise in the data. It is known that small changes in the assumed background in emittance data can have a large impact on the values of rms-emittances derived from the data[3]. The accuracy of area-emittance (total area-emittance of a given fraction of beam current) is also affected by incorrect background subtraction especially at large beam fractions. This report first discusses the difficulty of comparing rms and area-emittance values calculated by different analysis techniques employing different methods of background subtraction. Next the computer program used to analyze these data, is described and checked against analytic beam distributions characterized by known emittance values and containing various background currents. An analysis of a single data set is then carried out in detail to illustrate the technique and figures summarizing the characteristic emittance-intensity curve for each source are presented. Finally, a table is given summarizing the normalized emittance (RMS and 90%) for each source.

EMITTANCE MEASUREMENT AND ANALYSIS

Determination of the values of rms and area ion beam emittance from beam measurements is a difficult process potentially including systematic errors arising from both the experimental apparatus and data analysis. Popular emittance measurement devices include: the pepper pot (PP) which measures the 4-dimensional subspace {x, x', y, y'} of 6-dimensional phase space which the beam occupies; the slit and collector scanner (SCS) and the electric sweep scanner (ESS) both measuring transverse subspaces {x, x'} and {y, y'} independently[4]. All data analyzed in this report were measured using the two latter devices. Experimental problems include finite-slit dimensions, signal to noise ratio, aperture scattering, space charge transformation of the sample beamlet and contribution of secondary particles to the measured beamlet current[5].

Problems incurred during data analysis involve incorrect subtraction of background data that may be present, in measured data even after subtracting beamless scans. Typically, this effect arises from amplifier noise and drift[3]. In order to perform meaningful emittance measurements, the acceptance of the emittance device must be somewhat larger than the beam halo. This implies that when measuring the transverse subspaces ({x, x'}, {y, y'}), the majority of the measured data is background and is located at large distances from the origin. Thus, even a small bias on the current amplifier(s) or associated electronics can strongly affect the calculated rms emittances[3]. This is especially true for data points far from the origin since emittances are calculated from the second moments of the distribution and are weighted by x^2 and x'^2.

One of the principle difficulties in comparing emittances measured and analyzed in different laboratories is that each lab generally employs a different technique to separate real beam from background in measured data sets[6]. One popular technique involves tracking the number of data pixels, which fall within a particular interval of measured current intensity, and displaying this data as a histogram. The average background is then determined by finding the center of the dominant peak in the histogram. This average background is then subtracted from the overall data set and emittances are calculated.

Another approach involves calculating the rms-emittance of the entire data set as a function of different thresholding levels: i.e. levels of beam intensity below which data are ignored. One typically observes two distinct slopes in these plots: a region where calculated emittance depends strongly on the threshold parameter (steep slope) where the more abundant background data are being clipped and a region where the calculated emittance depends less strongly (shallower slope) on the threshold parameter, the beam region. In this approach, the emittance value corresponding to the intersection of these two regions is taken as the best measure of the rms-emittance.

A third approach is to threshold the data at different levels and plot the Twiss parameters α and β versus threshold value. As the threshold parameter is increased, the transition from clipping background to clipping beam is usually observed as rapidly changing values of α and β begin to assume slowly varying values characteristic of the beam. The threshold value used to separate beam from background is then determined from this transition region. The disadvantage of this technique is that significantly aberrated beam can have largely different Twiss parameters from the principle beam making it difficult to distinguish background from beam in the data set.

Still another approach, known as exclusion analysis, is employed by a number of computer programs where the user can graphically inspect data over the x, x' plane and define arbitrary boundaries beyond which data is ignored. This allows the user to effectively trim the data removing parts which appear to be background leaving what appears to be beam. The principle disadvantage of this technique is obvious: calculated rms-emittance values depend on the subjective judgment of the user rather than on formal systematics.

In general, these techniques can lead to very different values of rms-emittances that are then quoted in the literature. In Ref. 3, for example, the authors show that analysis of a single data set using these different approaches can lead to variation as large as ~50% in the calculated rms-emittance value. Also each of the above techniques of data reduction tends to underestimate the emittance by excluding or thresholding small current values, which are assumed to be background, but could, in fact, be real current from the beam halo. In the next sections we describe the computer program employed to analyze data and then develop the approach of elliptical exclusion analysis used to separate beam from background data in the present study.

THE EMITTANCE ANALYSIS PROGRAM

The program is a visual basic macro procedure that is used in conjunction with an Excel spreadsheet to provide a user-friendly computational platform on which a broad range of emittance analysis can be conducted. Raw emittance data, in the form of columns (x, x' and detector current i) are simply pasted into the spreadsheet and analysis is conducted through a 5-page user-form with each page corresponding to a major function of the program. Figure 1 shows the user interface of the program.

The only global variables used by the program are those arrays which store the emittance data: x, x' and i, and all other variables are defined locally within the individual subroutines to keep the programming highly modular. Each subroutine can be called from the user form and it modifies or performs calculations on the data as requested. At any time the user can request the Twiss parameters of the current distribution including β γ normalized and unnormalized rms-emittance, call for a 3D plot or 2D contour plot of the data over a specified window, center the data, refresh the data from the spreadsheet, or employ a data filter.

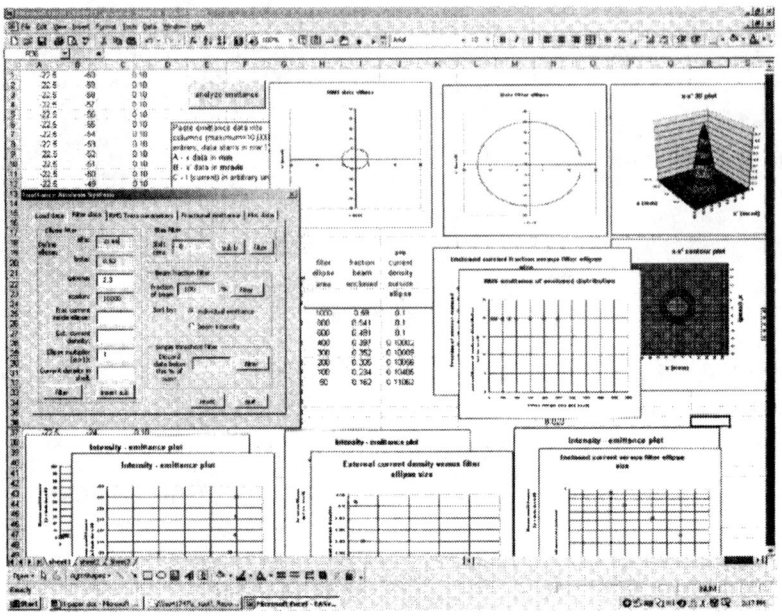

FIGURE 1. User interface used by the emittance analysis program. Data is simply pasted into the spreadsheet and analysis is controlled through a 5-page user-form. Graphics are displayed on the sheet along with numerical output tabulated in the form of many plots.

Data filters allow the user to do simple thresholding, shift the zero of the data, specify the fraction of the beam which undergoes analysis (e.g. the most intense 10% of the beam or 10% of the beam with the lowest emittance) or employ a data filter ellipse. This ellipse sets a user specified boundary on the x-x' plane which is elliptical in shape and allows exclusion of data points exterior to the ellipse. The program allows the user to specify the area, orientation and aspect ratio of the ellipse (ε, α, β) and the subroutine computes the fraction of the beam contained within the ellipse, the average current intensity of all data pixels external to the ellipse and then excludes these pixels from emittance analysis. An elliptical shell can also be specified where the average current intensity is determined within the shell. These techniques are used to perform elliptical exclusion analysis, which is described later in this paper.

Another subroutine calculates the area-emittance and fraction of beam enclosed within a given equal-intensity contour of the emittance data. This is used to produce plots of area-emittance for a given fraction of the beam. Automated plots are produced which include: the rms-ellipse and the filter ellipse on the x-x' plane; area-emittance and rms-emittance versus beam fraction and filter ellipse area; and external current density versus filter ellipse area. The function of these plots in data analysis will be described later in this paper. This program is available for distribution - contact the author for details[7].

ANALYTIC EMITTANCE DISTRIBUTIONS

Before analyzing emittance distributions from ion sources the code is validated using a 2D Gaussian distribution and a uniform ellipse (2D projection of the 4D Kapchinskii-Vladimirskii distribution). A uniform background of varying fractional amounts was added to each distribution to verify that the code and analysis technique could adequately distinguish background. Eq. 1 shows the Gaussian distribution which was input into the emittance analysis program distributed over ~5000 data points.

$$f(x,x') = \frac{1}{2\pi\sigma\sigma'} \exp\left(-\frac{1}{2}\left(\frac{x^2}{\sigma^2} + \frac{x'^2}{\sigma'^2}\right)\right) + b \quad \sigma = \sigma' = 5 \tag{1}$$

Here σ and σ' are the variances of the distribution with respect to x and x' and b is the specified background level. Using the standard definition of rms-emittance we can show analytically that a 2D Gaussian distribution has rms-emittance values of

$$\varepsilon_{rms} = \sqrt{\langle x^2 \rangle \langle x'^2 \rangle - \langle xx' \rangle^2} = \sigma\sigma' = 25. \tag{2}$$

Here the $\langle \rangle$ brackets have their usual meaning:

$$\langle a \rangle = \sum_x \sum_{x'} a\, f(x,x') dx\, dx' \tag{3}$$

Elliptical exclusion analysis, described in the next section, was applied to this distribution using the emittance program and the expected result of 25.00 was obtained for any background levels b. The area-emittance ε of a given beam fraction F can be obtained by integrating Eq. 1:

$$\varepsilon(F) = 2\varepsilon_{rms} \ln\left(\frac{1}{1-F}\right). \tag{4}$$

Fig. 2 shows a plot of area-emittance versus beam fraction (plotted as ln(1/(1-F))). The line represents Eq. 4 while the data points are output from the analysis program. Excellent agreement is clearly observed between the analytic and numerical results.

Similar analysis was performed using a uniform ellipse with an area of 100 π mm mrad and various background levels distributed over ~5000 data points. Elliptical exclusion analysis was applied and the result of 25.00 was obtained in agreement with the expected result of $4\varepsilon_{rms}^5$ for this distribution. This result held for different ellipse orientations and aspect ratios as well as with different background levels.

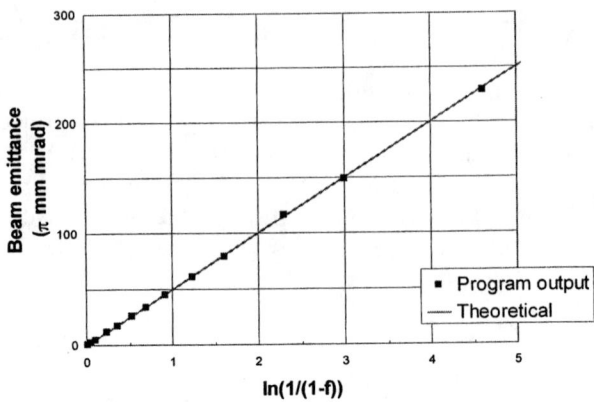

FIGURE 2. A plot of area-emittance versus beam fraction plotted as ln(1/(1-F)). Excellent agreement is observed between the analytical and numerical results.

As a further validation of the code, 5 different data sets were analyzed both by the LBNL emittance code[8] and the present code and exact agreement was observed in each case.

METHOD OF DATA ANALYSIS

The technique of elliptical exclusion analysis (SCUBEEx) is fully developed in Ref. 3 and will only be summarized here. It is based on the hypothesis that an ellipse in 2D phase space can be defined around the beam which contains essentially all the real beam current and data pixels exterior to the ellipse are background. The hypothesis is confirmed, on a case-by-case basis, by showing that once the average background, determined from data outside the ellipse, is subtracted from data inside the ellipse, the rms-emittance calculated from data within the ellipse remains essentially invariant as the ellipse is varied in size over a reasonable range. This approach is computationally more challenging than the approaches discussed earlier, but is fundamentally sounder.

Using the emittance analysis program, the procedure followed is straightforward: The orientation and aspect ratio of the ellipse are determined by defining the Twiss parameters, α and β to correspond with the principle beam data (thresholded at large values ~10%) and centered on the data. The ellipse area (= $\pi \varepsilon$), starting from large values, is reduced while tracking average current in pixels exterior to the ellipse. This value is then subtracted from current in all pixels inside the ellipse and the rms-emittance is calculated using only the data inside the ellipse. Plots of this ellipse area versus rms-emittance calculated in this way clearly show stable emittance values over a large range of 'filter' ellipse areas. All of the data sets analyzed in the preparation of

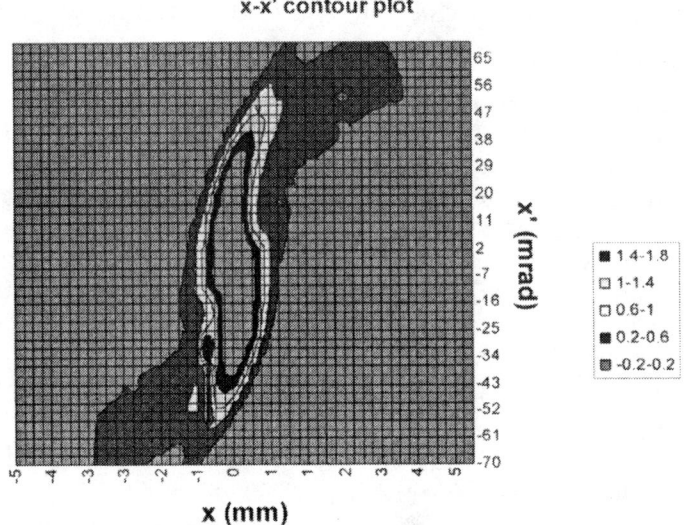

FIGURE 3. Raw emittance data from the LBNL ion source and LEBT in the vertical direction for 30mA of H- at 65kV extracted from the source through a ϕ = 7 mm circular aperture. Contours are drawn through the data at 20% intervals with respect to beam intensity.

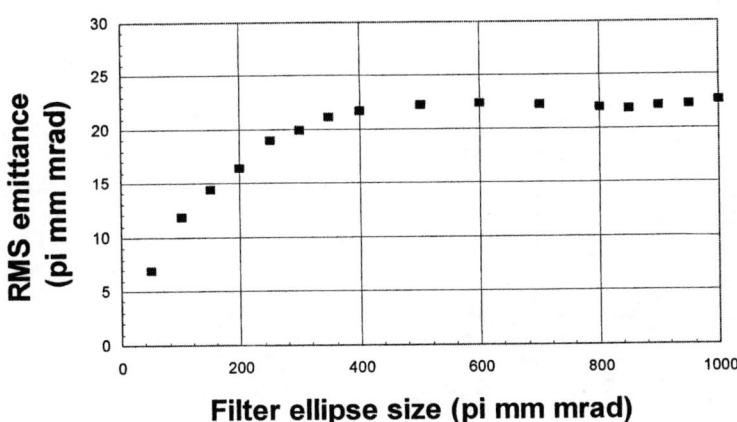

FIGURE 4. rms-emittance calculated inside the filter ellipse after subtracting background determined from data points outside the filter ellipse for different filter ellipse areas.

this report showed a nearly flat plateau when the filter ellipse was larger than the beam current, which fell off precipitously as real beam was excluded from the emittance calculation.

A detailed description of the analysis of one data set will now be provided to illustrate the technique. The results from analysis of the other data sets will be presented in summary form in the last section of this report. Fig. 3 shows the raw data obtained from the LBNL multicusp source and electrostatic LEBT (Low Energy Beam Transport). Fig. 4 shows the results of performing elliptical exclusion analysis on this data. The plot shows the rms-emittance (unnormalized) calculated inside a filter ellipse after subtracting average background current computed from pixels exterior to the ellipse. The graph shows this emittance for different values of the filter ellipse area. In this case, the RMS emittance is found to be 22 ± 0.5 π mm mrad with the uncertainty given by the variance of the emittance values over a range of ellipse areas of 400-1000 π mm mrad. The variance quoted above represents the uncertainty only with respect to this analysis approach (uncertainty of background level) and by no means represents the overall experimental uncertainty of the emittance measurement.

Plots of emittance versus fraction of beam are generated using the following procedure: the orientation and aspect ratio of the filter ellipse (α, β) is set as in exclusion analysis and the filter ellipse is once again reduced in area about the center of the data. The average background current per pixel exterior to the ellipse is then plotted against ellipse size as shown in Fig. 5. Inspection of this graph allows easy determination of the beam boundary, which, in this case is 450 π mm mrad and the average background subtracted from the data is 0.0111 or 0.16% of the maximum pixel current. Occasionally, a clear transition from beam to background is not observed and a better choice of α and β is needed. The code provides a means of manually setting these values based on inspection of the data plots.

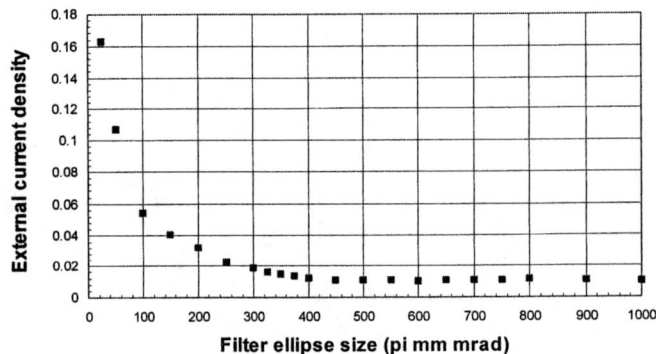

FIGURE 5. Average current density of data pixels located outside the filter ellipse as it is reduced in area about the center of the data.

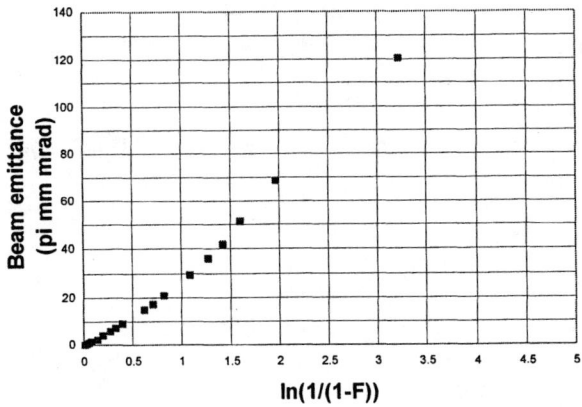

FIGURE 6. Total emittance of the beam plotted against fraction of the beam expressed as ln(1/(1-F)).

Once the data is reduced with the filter ellipse and background current is subtracted, the program automatically generates plots of the total emittance contained within a contour of equal current density plotted as a fraction of beam. Since most emittance data tends to be close to Gaussian, the program plots data in the form: total emittance versus ln(1/(1-F)) where F is the fraction of beam contained within the corresponding emittance. Plots of this type tend to delineate departure from Gaussian behavior. Fig. 6 shows this plot for the current data set and shows the emittance of 90% of the beam is 80 π mm mrad or 0.94 π mm mrad βγ normalized.

ION SOURCES INVESTIGATED

As mentioned above, laboratories which employ high-brightness H⁻ ion sources for accelerator application were asked to contribute raw emittance data measured as close in the beam-line to the source as possible and under conditions required for nominal accelerator operation: not conditions optimized for minimum emittance from the source. In some cases measurements were made on an off-line test stand. Thus, this data summary features typical rather than optimal emittance characteristics of each type of source and does not show a comparison between the best possible performance of each source and LEBT design.

LBNL Multi-cusp Volume Source and Electrostatic LEBT

This ion source and LEBT developed for the US Spallation Neutron Source (SNS) were designed and built by LBNL. Details of the ion source and LEBT can be found in the literature[9][10] along with details of the emittance measurements[8], which were performed using a ESS system. Transverse emittance was measured several mm downstream of a double Einzel lens electrostatic LEBT. A total of 4 data sets were analyzed: x and y for the early LEBT design (exit electrode: ϕ = 11 mm) and final LEBT design (exit electrode: ϕ = 7 mm). The source has been operated at extraction voltages of 65 kV producing H⁻ beam from 20-50 mA.

ISIS Penning Source

Details of the ion source[11], used as the workhorse of the ISIS neutron and muon facility, and details of the emittance measurements can be found in the literature[12]. Briefly, the ion source is a Penning-type, cesium seeded H⁻, surface plasma source. Ions are extracted from the plasma through a slit 0.6 x 10 mm, at 17 kV, passed through a 90° sector magnet then accelerated to 35kV and passed through two lenses of a magnetic LEBT to an SCS emittance detector ~ 690 mm from the source. These measurements were performed on the RFQ test stand[13] where ~30 mA of H⁻ ions were transmitted to the emittance detector through a small aperture gas baffle as in the case of the ISIS injector[12]. This source can produce up to ~50 mA.

BNL Magnetron Source

The BNL magnetron source has, for many years, been the workhorse of the 200 MeV linac, which injects the AGS booster. Details of the ion source and details of the emittance measurements can be found in the literature[14]. Briefly, the ion source is a magnetron-type, surface plasma source operated with cesium. Ions are extracted from the plasma through a 2.8 mm diameter aperture, at 35 kV, where they drift ~15 cm and are passed to a SCS emittance detector. Data sets of 65 and 100 mA were analyzed, 100 mA being close to the maximum output of this type of source.

LANL Surface Conversion Source and Magnetic LEBT

The LANSCE ion source is a filament-driven, multicusp, cesium seeded, surface conversion source originally designed by LBL and modified at LANL[15]. In its present form, the operations source has been the workhorse of this facility for many years. H⁻ ions are primarily formed on a converter surface by H⁺ bombardment and accelerated towards the extraction opening by a ~250 V potential. The ions then exit the discharge chamber through a 13 mm diameter aperture; enter an 80 kV acceleration column; undergo focusing by a single solenoid lens and are detected using a SCS emittance device. Approximately 15 mA of H⁻ ions were extracted from the data set analyzed although 20 mA is close to the maximum output of the source.

ANALYSIS RESULTS

Table I shows a summary of the emittance values derived from each data set analyzed in this study. All of the emittance values shown are $\beta\gamma$ normalized. The 5th column contains rms-emittances defined by Eq. 2. The 6th column contains the area-emittance defined as the total area in the x-x' plane bounded by an equal intensity contour containing 90% of the total beam. Conditions of the measurement are also given in the table. Beware that apparently similar emittance definitions used in the literature. For example, Ref. 2 defines emittance $\varepsilon_n^{90\%}$ as described above while Ref. 16 Defines $\varepsilon_n^{90\%}$ as the area enclosed by the equal intensity contour defined to be 90% of the maximum current in a single pixel. Fig. 7 shows a plot giving the total emittance of the beam as a function of beam fraction F plotted as $\ln(1/(1-F))$ to facilitate comparison with Gaussian behavior. Emittance plots of this type are quite useful since, to first order, this curve remains invariant under propagation from the source [5] and provides the intrinsic limit of the maximum H$^-$ current an ion source can deliver into a given acceptance provided proper matching can be achieved. For cases which exhibited a high degree of Gaussian behavior, a least squares fit was applied to the data and the slope of the resulting line was used to cross check the value of ε_n^{rms} using Eq. 4. Excellent agreement was obtained in each case. Proper exclusion of background current is important to accurately calculate emittance values as F→1.

TABLE I. Emittance values extracted from the analyzed data sets.

Ion source	LEBT	Beam energy (keV)	Beam current (mA)	ε_n^{rms} (π mm mrad)	$\varepsilon_n^{90\%}$ (π mm mrad)
LBNL-X	Electrostatic 13 mm exit	65	30	0.27	1.0
LBNL-Y	Electrostatic 13 mm exit	65	30	0.26	1.0
LBNL-X	Electrostatic 7 mm exit	65	33	0.22	0.82
LBNL-Y	Electrostatic 7 mm exit	65	33	0.18	0.65
ISIS-X	2 magnetic lenses	35	30	0.19	0.65
ISIS-Y	2 magnetic lenses	35	30	0.20	0.85
BNL	None	35	65	0.33	1.6
BNL	None	40	100	0.56	1.9
LANL-Y	1 magnetic lens	80	15	0.18	0.73

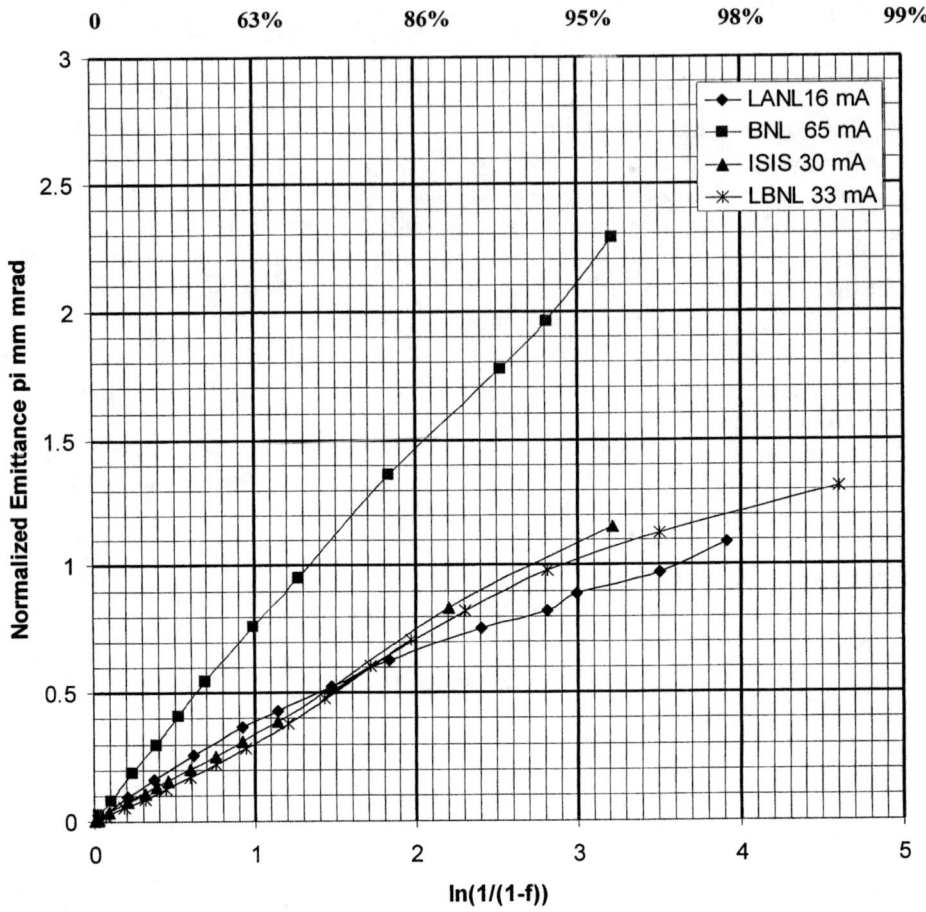

FIGURE 7. Normalized area-emittance for different fractions of total beam current plotted as $\ln(1/(1-F))$. Corresponding percentages of total beam are plotted on the upper horizontal axis. In cases where two transverse emittance scans were analyzed, the larger emittance is plotted here. LBNL data is plotted for a LEBT exit aperture of 7 mm.

CONCLUSION

Software has been developed which allows unbiased, self-consistent elliptical exclusion analysis (SCUBEEx) to be applied in a straightforward manner to arbitrary emittance data sets. The program was checked against known analytical distributions and excellent agreement was observed. This data analysis technique was then applied to raw data measured from each major type of high-brightness H$^-$ ion source: Penning, magnetron, RF-multicusp and surface conversion multicusp. In each case, good beam and background separation was achieved as shown in Fig. 5. The LBNL and ISIS sources had similar beam brightness, ~ 6 mA/(mm mrad)^2, defined in terms of the $\varepsilon_n^{90\%}$ emittance values. The LANL and BNL sources also had similar brightness ~ 2.7 mA/(mm mrad)^2 but were lower than the other sources. This result was expected since the surface produced H$^-$ are known to be ejected from the surface in broad energy-angular distributions diluting the brightness of the source[17]. Plots of area emittance versus beam fraction (Fig. 7) are generally conserved during beam transport and are therefore characteristic of the ion source. In all cases, a near Gaussian behavior is observed. Curves in this form are useful since they give the maximum possible current an ion source can deliver into a given acceptance under ideal matching conditions.

ACKNOWLEDGEMENTS

The authors wish to thank the technicians, engineers and scientists, from LBNL, LANL, RAL-ISIS and BNL who have made the high quality measurements upon which this article is based.

REFERENCES

[1] Dimov, G.I., *Rev. of Sci. Instrum.* **67**, 3393-3404 (1996).
[2] Alonso, J.R., *Rev. of Sci. Instrum.* **67**, 1308-1313 (1996).
[3] Stockli, M.P. et al., *these proceedings (2002)*
[4] Allison, P., *Proceedings of the Forth International Symposium on the Production and Neutralization of Negative Ions and Beams,* Brookhaven, NY, AIP press, p 465-481 (1986).
[5] LeJeune, C., and Aubert, J., in *Applied Charged Particle Optics*, edited by A. Septier, Academic Press, New York, 1980, pp. 159-259.
[6] A private communication: ORNL, LBNL, DESY, ISIS, CERN, BNL, FNAL and LANL.
[7] Email: welton@ornl.gov.
[8] Thomae, R., et al, *Rev. of Sci. Instrum.* **71**, 1213-1215 (2000).
[9] Keller, R., et al, *Rev. of Sci. Instrum.* **73**, 914 (2002).
[10] Keller, R., et al, *these proceedings (2002)*.
[11] Thomason, J.W.G., et al, *Rev. of Sci. Instrum.* **73**, 896-898 (2002).
[12] Letchford, A.P. et al, *Proccedings of the 8th European Particle Accelerator Conf. Paris 2002* contribution # THPLE045.
[13] Thomason, J.W.G., et al, *Proccedings of the 8th European Particle Accelerator Conf. Paris 2002* contribution # THPR1011.

[14] Alessi, J., *20th ICFA Advanced Beam Dynamics Workshop on High Intensity High Brightness Hadron Beams,* Fermilab April 8 - 12, 2002.
[15] York, R.L. and Stevens, R.R., IEEE Trans. Nucl. Science, Vol. **NS-30**, 4-8 (1983).
[16] Peters, J., *Rev. of Sci. Instrum.* **71**, 1069 (2000).
[17] Welton, RF, "Overview of High Brightness H⁻ Ion Sources", *the 21ˢᵗ International Linear Accelerator Conference, Gyeongju, Korea, 2002.*

NEW CONCEPTS

The CEA/Saclay 2.45 GHz Microwave Ion Source for H$^-$ Ion Production.

R. Gobin, K. Benmeziane, O. Delferrière, R. Ferdinand, F. Harrault, and J. D. Sherman*,

CEA/Saclay, DSM/DAPNIA, 91191 Gif/Yvette, France
* LANL, Los Alamos, N.M. 87 545, USA

Abstract. H$^-$ ions have now been observed in currents extracted from the CEA/Saclay ECR proton source. A pulsed, 90 degree dipole magnet has been installed 14.8 cm downstream from an ion beam extraction system designed to operate up to 10 kV. The plasma chamber can be biased by a negative or positive HV power supply and the diagnostics are grounded. The magnetic field is reduced to 200 G at the emission aperture. The measured positive hydrogen ion fractions show a large amount of H$_2^+$. A 32 wire profiler has also been installed at the dipole magnet location to analyze the total extracted beam and compare it with calculations. Profile measurements at negative extraction potential indicate an important deflection of the beam when the magnetic filter is installed. To measure a consistent electron to H- ratio, the magnetic filter has been tuned in order to avoid steering the beam off axis. Preliminary results will be reported as well as analysis of the profile.

This work is supported by the European Commission under contract n°: HPRI-CT-2001-50021

INTRODUCTION

Applications of high current accelerators include the production of high flux neutron beams for spallation reactions [1] and neutrino production for high-energy particle physics [2]. The high intensity beams for these accelerators may reach multi-GeV energies. In France, CEA and CNRS have undertaken an important R&D program on very high beam power (MW class) light-ion accelerators for several years. Part of the R&D efforts is concentrated on the High Intensity Proton Injector (IPHI) [3] demonstrator project. The High Intensity Light Ion Source (SILHI) development, based on the 2.45 GHz ECR plasma production, has been performed for several years leading to valuable experience in high current proton beam production [4]. Taking this experience into account, CEA decided to develop a hydrogen negative ion source based on the ECR plasma production for possible application in future H$^-$ projects such as ESS.
In this report, the first experimental set-up is briefly reported in section 1. Preliminary plasma analysis are presented in section 2. The magnetic configuration has a large influence on the extracted beam measurements. Results for the important e/H$^-$ ratio appear to be strongly related to the magnetic filter configuration, and will be discussed in the third section. The installation of a dipole magnet with Faraday cup has resulted in the definite identification of H$^-$ ions extracted from the microwave-generated plasma.

EXPERIMENTAL SET-UP

Taking into account the good experience obtained in high intensity beam production with the SILHI source, CEA has decided to develop a new negative hydrogen ion source. This source is also based on the ECR plasma generation operating at 2.45 GHz. Two coils are used to provide the B_{ECR} = 875 Gauss axial magnetic field. A protected quartz window separates the standard WR 284 rectangular waveguide plasma chamber from the 1.2 kW magnetron RF source. A three ridged transition, located at the plasma chamber entrance, allows concentrated RF field on the source axis. The water cooled plasma chamber is made of copper. The 210 mm plasma chamber length has been chosen to limit as much as possible the axial magnetic field close to the extraction zone. To avoid high energy electrons in this area, a tunable C-shape magnetic filter (MF) separates the H$^-$ production zone from the ECR plasma generator. A second C-shape magnetic dipole (Sep) is installed in the diagnostic box to force electron dumping on the extraction electrode.

The first step is to verify the effective H$^-$ ion production as demonstrated elsewhere [5, 6]. To facilitate tuning and rapid configuration changes, computations have been done to fix the maximum beam energy at 10 kV (Axcel and Opera codes). The 5 mm diameter plasma electrode, the plasma chamber and the source ancillaries were initially grounded. Only the extraction electrode and the collector were insulated to operate at high voltage. To avoid source and collector heating, the magnetron is operating in pulsed mode (typically 1 ms at 10 or 20 Hz).

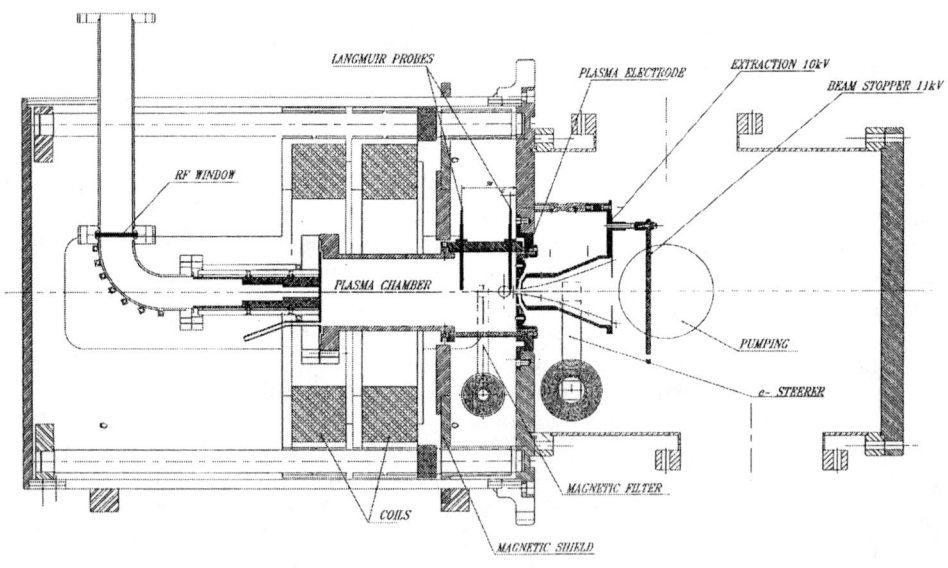

FIGURE 1. Cross-sectional view of the experimental set-up.

PLASMA ANALYSIS

Generally, the accelerator source community focuses work on the extracted beam characteristics. Presently, to better understand fundamental source behavior, diagnostic ports installed on the plasma chamber allow plasma analysis. Either two Langmuir probes or a monochromator viewport can give information on the plasma. Figure 1 shows the first experimental set-up with two Langmuir probes. After removing probes, plasma emission spectroscopy analysis indicated specific lines showing presence of vibrationally excited H_2 molecules which move to the H^- production region.

Langmuir probe measurements

Measurements have been performed with a 1.7x1 mm rectangular probe in both the driver region (58 mm from the plasma electrode aperture) and the production region (6 mm from the aperture). At the second location, the axial magnetic field was lower than 100 Gauss. The MF with 10 mm distance pole face (located 12 mm from the emission aperture) provides a transverse field integral of about 600 G.cm. Figure 2 (in semi log scale) represents the probe measurements plotted for different magnetic configurations of MF and Sep. The derived electron temperature is 6.7 eV in the driver region independent of the MF and Sep current settings. However, in the production region the derived electron temperatures range from 3.5 to 5.3 eV depending on the FM and Sep current excitations.

Spectroscopy analysis

Plasma spectroscopy has been made using a Jobin-Yvon monochromator equipped with a 300-1000 nm grating. For this experiment, a view port replaced the probe support flange and MF was moved back to 40 mm from the emission aperture. Specific lines (λ = 715.3 nm, 674.0 nm, 631.5 nm, 627.0 nm) appeared when MF is switched "on" (Fig. 3). This indicates the MF influence on the plasma behavior.

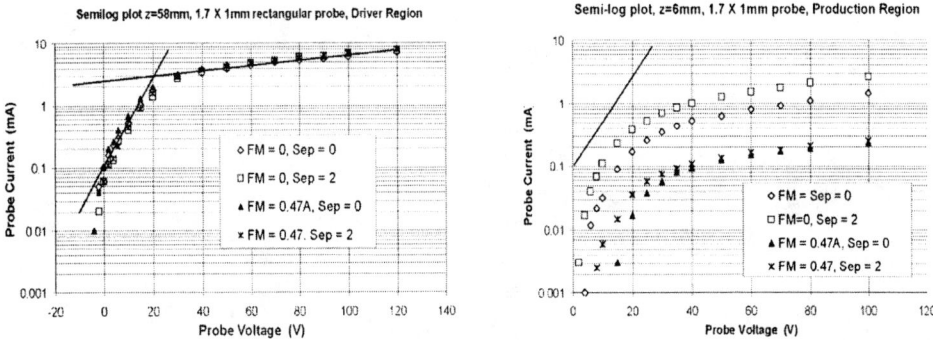

Figure 2: Langmuir probe measurements in the driver region (left) and in the production region (right). The solid line in the right figure is a reproduction of the solid line used to extract the electron temperature in the driver region.

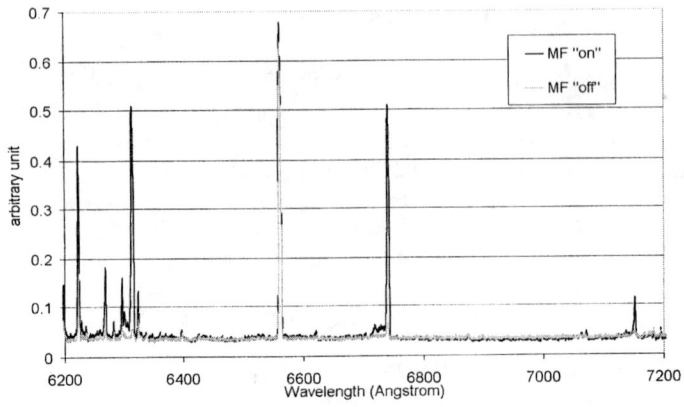

Figure 3: Plasma spectrum comparison : MF off (grey), MF on (black)

EXTRACTED BEAM ANALYSIS

Experimental set-up changes

The first extracted beam measurements did not definitely reveal H⁻ ion production [7]. To improve the diagnostic efficiency, the plasma chamber is now insulated and could be biased by negative or positive power supplies up to 10 kV. Sep (electron separator) has been removed and in the same time, a grounded 90° dipole analyzer magnet (DAM), located behind the main collector allows beamlet analysis on the mechanical apparatus axis [8]. The results are measured on an electrically shielded Faraday cup as a function of the DAM excitation.

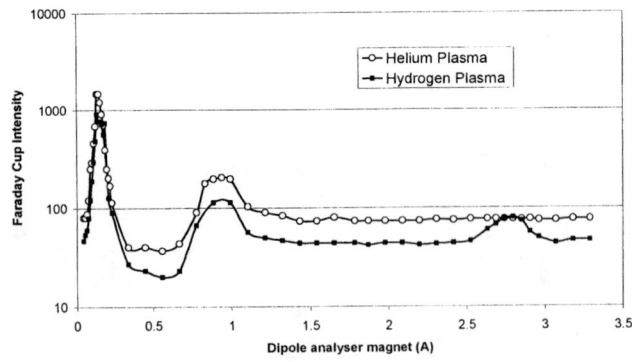

FIGURE 4. Both He and H extracted beam comparison vs analyzer magnet excitation. The peak located at 2.8 A dipole excitation, confirms the presence of H- ions.

To calibrate the analyzer, previous measurements were made with positive ion beam. The source was fed with hydrogen gas and the plasma chamber was set at 6.4 kV. A 3 mA total extracted beam has been produced with the following species fraction: $H^+ = 23\%$, $H_2^+ = 47\%$ and $H_3^+ = 30\%$. The H^+ peak current reaches a maximum level when the DAM is set to 2.8 A. The proton fraction ratio shows hydrogen dissociation process.

Negative charge extraction

To verify the H^- production, the plasma chamber is set at -6.4 kV. Total 40 mA beams are extracted. A large peak rises for a low DAM excitation (0.15 A) which is attributed to electron currents. Another peak grows when DAM excitation arrives at 2.8 A and, following the expected results derived from H^+ positive charge analysis, this peak is assigned to H^- currents. An unidentified peak occurs at 1-A DAM excitation. The unknown peak is believed to arise from electron currents. An improved collimation system for the Faraday cup entrance may eliminate this signal. To confirm the H^- measurement, helium gas has been injected in the source and the dipole analysis revealed only electron peaks at 0.15 and 1,0 A, and no peak appeared at 2.8 A DAM excitation (Fig.4). By comparing electrons and H^- height signals, the e^-/H^- ratio ranges from 11 to 27 which could lead to a total H^- ion production close to 1 mA.

In addition, two other sets of e^- and H^- data were acquired at 3.2 and 9.0 keV beam energy. The peak displacements are in good agreement with momentum scaling predictions (Fig 5). The unknown peak seems to scale up in $B\rho$ as the extraction voltage increases.

The low Faraday cup signal forces us to better understand the beam transport. So a movable 32 wire profiler equipped with appropriate amplifiers now replaces DAM. The main collector with a 6 mm DAM entrance hole is replaced by an insulated plate with a 30mm diameter hole, defining the profiler entrance aperture. Profile measurements (Fig. 6 left) indicate a strong deflection of the negative beam in the conditions of the first experiment (with MF excitation = 0 A). By tuning the MF (1.2 A), the deflection decreases (Fig 6 right).

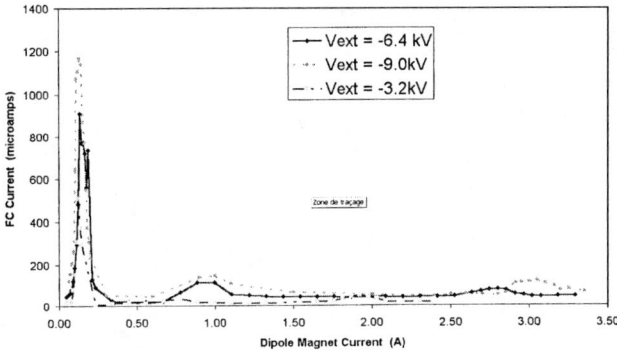

Figure 5: Faraday cup measurements vs DAM excitation for 3.2, 6.4 and 9.0 keV extracted beams.

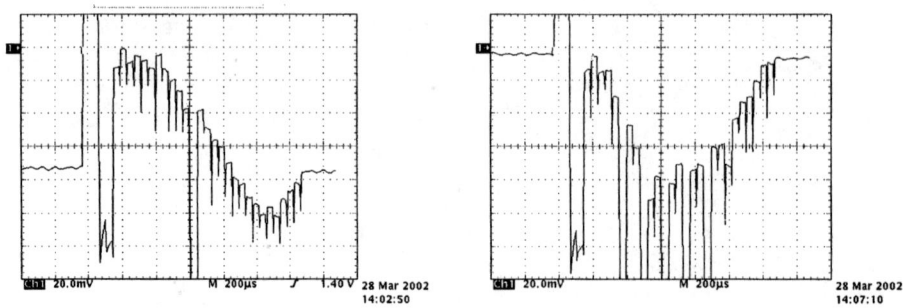

Figure 6: Profile measurement showing the MF excitation influence on beam centering (left: MF = 0 A and right: MF = 1.2 A)

This shows the strong influence of the induced transverse magnetic field in the MF iron yoke previously measured (Fig. 7 left). By totally removing the MF, the extracted beam was only slightly deflected and thus confirmed this influence (Fig 7 right). At the same time, the total extracted current increased from 30 mA to 37 mA.

Recently a new experiment has been performed to analyze the total extracted beam and to verify the e-/H⁻ ratio. A large Faraday cup equipped with secondary electron suppressor allows the measurement of the total beam crossing the large collector with 30mm entrance aperture. In addition electron and ion separation is done by Sep located 80mm from the extraction aperture. MF is still removed. Measurements show a low amount of H⁻ ions (Fig.8 left) compared to the total signal mainly due to electrons (Fig.8 right) leading to e-/H⁻ ratio as high as 8000. It is noted a "afterglow" peak occurs for the negative ion signal. The same "afterglow" signal was observed on the Faraday cup installed behind DAM.

The first step to demonstrate the production of H⁻ ions with an ECR plasma generated source has been made. The reported measurements are powerful evidence that H⁻ ion have been formed and extracted from such a kind of source:
- the mass scaling between H^+ and H^- is correct

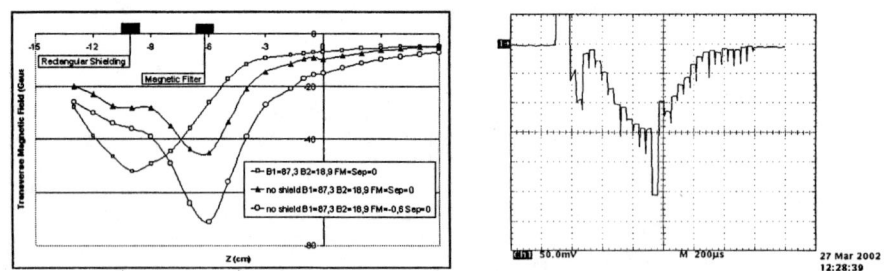

Figure 7: Transverse magnetic field with MF (left) and Profile with MF removed (right).

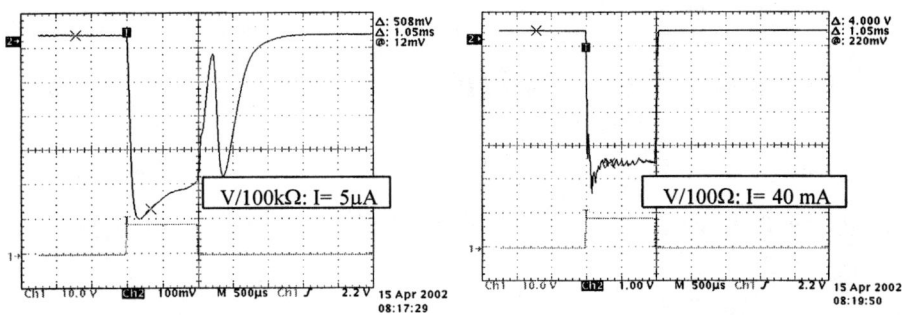

Figure 8: Faraday cup total beam analysis with Sep (left: H- ions; right total beam, mainly electrons)

- momentum scaling is correct
- no peak at H⁻ location with plasma helium

Moreover, the e-/H⁻ ratio strongly varies as a function of the magnetic configuration of the apparatus. As the MF is not yet reinstalled, the high e-/H⁻ ratio measured in the last experiment may be due to MF absence.

In the near future, MF will be reinstalled. Then, new measurements with hydrogen and helium plasma will be planned to definitely fix the e-/H⁻ ratio.

ACKNOWLEDGMENTS

Many thanks to the members of the IPHI team for their contributions, especially to G. Charruau and Y. Gauthier for their technical participation. The authors would also thank M. Bacal, J. Faure, G. Gousset, A. Girard, C. Jacquot, G. Melin, for their fruitful collaboration and valuable discussions.

REFERENCES

1 Duperrier R., et. al., "The ESS Front End Associated with the SC Linac", ESS TAC Meeting no 1, FZ Juelich, January 7-9, 2002
2 Garoby R., "Status & plans of the SPL study at CERN", *ICFA 2002 proceedings*, Fermilab, 2002.
3 Beauvais P-Y., "Status report on the construction of the French High-Intensity Proton Injector (IPHI)", *EPAC Conference proceedings*, Paris, France, 2002
4 Gobin R. et al., "Saclay High Intensity Light Ion Source Status", *EPAC Conference proceedings*, Paris, France, 2002
5 Tanaka M. and Amemiya K., *Rev. of Sci. Instrum.* **71**, 1125-1127 (2000).
6 Jayamanna K. et. al., *Rev. Sci. Instrum.* **67**, 1061-1063 (1996).
7 Gobin R., et al., *Rev. Sci. Instrum.* **73**, 983-985 (2002).
8 Gobin R. et al., "Observation of H- Ions Extracted from a 2.45 GHz Microwave Ion Source", *EPAC Conference proceedings*, Paris, France, 2002

First Simulations of the Cadarache SINGAP Experiments

HPL de Esch, D Boilson, R S Hemsworth, P Massmann and L Svensson

Association EURATOM - CEA CADARACHE, DRFC/SCCP
13108 ST PAUL LEZ DURANCE Cedex, France

Abstract. The SINGAP (SINgle APerture - SINgle GAP) acceleration concept is a simplified alternative to the multi-aperture, multi-grid acceleration of the ITER Neutral Beam reference design. Our project aims to demonstrate reliable multi-second acceleration of a D⁻ beam to 1 MeV (~100 mA, 200 A/m^2), relevant to the ITER Neutral Beam Injection requirements and to validate the predictions of the simulation codes used to design the SINGAP accelerator for ITER.
The present campaign achieved (911 keV, 30A/m^2) and (600 keV, 60 A/m^2) D⁻ beams. The highest space charge effects have been obtained with 400 keV, 50 A/m^2 D⁻, which, although still a factor of 2.5 below the ITER value for space charge, is beginning to test the space charge aspects of the codes. Simulation results are compared with the experimental data for a variety of cases.

INTRODUCTION

The European concept for a 1 MeV, 40A negative ion based accelerator for the neutral beam injection system on ITER, the SINgle GAP, SINgle APerture (**SINGAP**) accelerator is an attractive alternative to the reference design, the so-called Multi-Aperture, Multi-Grid (**MAMuG**) accelerator, which is being developed in Japan. Where MAMuG requires 7 multi-aperture grids, each inclined differently in two directions for geometric beam steering, SINGAP proposes acceleration of a pre-accelerated beam in one single step to 1 MeV [1] (using 3 flat multi-aperture grids without needing geometric steering). The fourth (SINGAP) grid is a structure containing only 16 rectangular apertures, each of which accelerates a group of 80 beamlets to 1 MeV. The Cadarache experiments aim to validate this concept. A physics design for an ITER SINGAP accelerator was presented in [1].

In the second half of 2001, experiments were conducted with the 1 MV SINGAP installation at Cadarache, which uses a 1 MV Cockroft-Walton power supply with a maximum output current I_{max} of 100 mA [2]. H⁻ or D⁻ beams were extracted from the "sourcette" ion source [3] through 11 apertures and pre-accelerated to energies in the range of 12 to 50 keV. The post-accelerated beam is stopped on a graphite target, which allows determining beam currents and power density profiles by means of infrared calorimetry. The current densities measured calorimetrically were up to 6 mA/cm^2. The area of one aperture on the plasma grid is 1 cm^2, which means

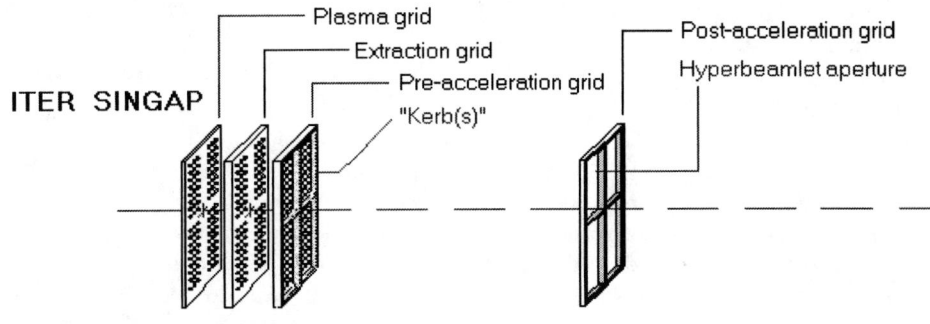
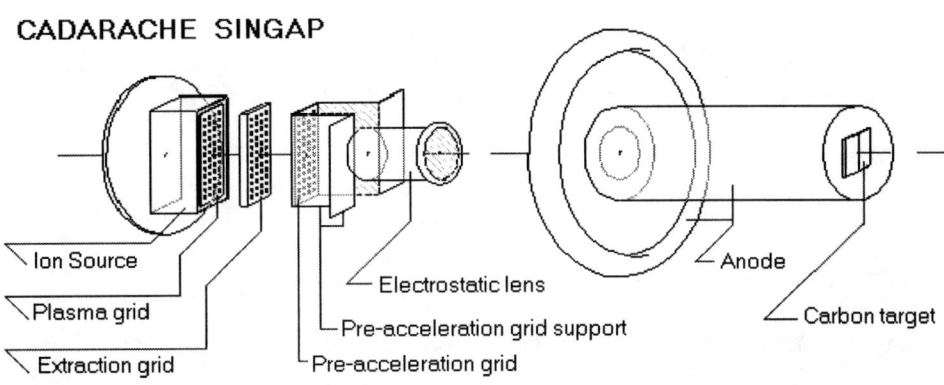

FIGURE 1. The ITER-SINGAP design and the Cadarache SINGAP experiment. Note that the Cadarache experiment presently features a long "kerb" formed by the pre-accelerator grid support and electrostatic lens. To achieve 11 beamlets, most apertures in the plasma grid have been blanked off, leaving only 11 apertures open.

that the total accelerated current is 11 times the current density. The pre-accelerated beam was post-accelerated in one single step to energies chosen between 300 and 914 keV. Due to an R=1 MΩ series resistor for its protection, the power supply is effectively limited to $V < 10^6$-IR, thus 900 kV at 100 mA.

Due to all this, the current density is subject to an upper limit of 9 mA/cm^2. We ran a maximum of 6 mA/cm^2. The source, which is poorly cooled and therefore short-pulse, can supply more than this, but having no spare we chose not to run the risk of breaking it.

In 2002, the post-acceleration gap was shortened from 625 mm to the ITER relevant 350 mm.

The purpose of this paper is to compare the experimental data with simulations. Bucalossi et al [2] have done the comparison for cases where the space charge of the beam was low (J~2 mA/cm^2) and the acceleration gap was 625 mm. They found reasonable agreement. During the 2001 campaign, this agreement was recovered at the

15% level **for low space charge cases**. Here, we compare in more detail and also test the modelling assumptions for higher space charge cases.

For ITER it is necessary to obtain 20 mA/cm^2 D$^-$ after stripping losses. In our test stand, a maximum of 6 mA/cm^2 D$^-$ was obtained at E=580 kV and a maximum of 5 mA/cm^2 D$^-$ was achieved at E=400 kV. Both cases are equivalent to a space charge that by virtue of its $1/\sqrt{E}$ dependence corresponds to J=8 mA/cm^2 at 1 MeV. Regarding space charge, the 2001 experiments achieved 40% of the ITER value and the experiments should show some space charge effects.

In contrast to the ITER design, the pre-accelerator grid supports of the Cadarache experiment form an electrostatic equipotential drift space before post acceleration takes place. The situation is as if the pre-accelerator were fitted with a 349 mm long "kerb" (the ITER kerb is foreseen to be only 35 mm). Included in the 349 mm is an electrostatic lens extension that provides cylindrical symmetry.

The situation is sketched in figure 1. The difference with ITER means that the present experiments are aimed at demonstrating the voltage holding of the main acceleration gap (350 mm) and validating the codes and working hypotheses used for the ITER SINGAP design.

MODELLING PROCEDURE

The simulation is carried out as follows:

a) Simulate the pre-acceleration stage with the CADSLAC code

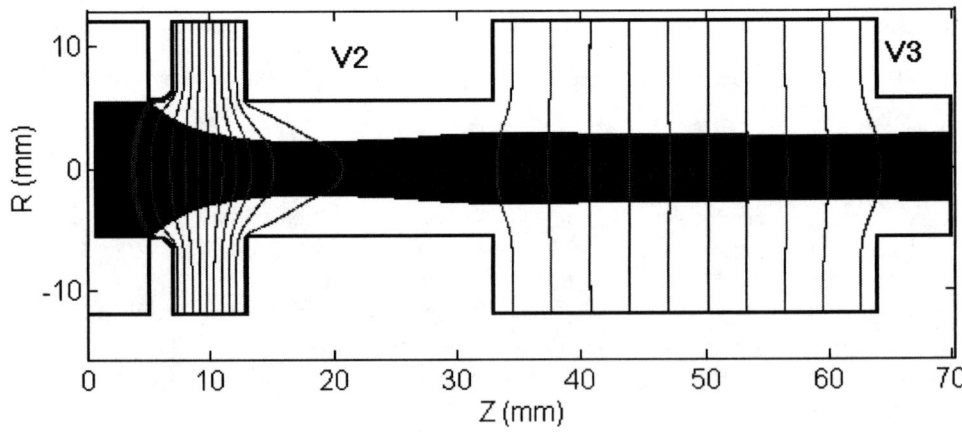

FIGURE 2: Example of a CADSLAC run with the actual geometry of the pre-accelerator.
The example is for shot 3332, where a 70 mA beam is extracted with V_2=4.4 kV and pre-accelerated to V_3=43 keV. (J_{ext}=6.2 mA/cm^2 D$^-$, 20% stripping losses result in J_{acc}=5 mA/cm^2). The calculated divergence of the resulting 43 keV beamlets is 10 mrad

The geometry of the pre-accelerator, the extraction voltage V_2, the pre-acceleration voltage V_3, the extracted current density J and the beam species (either H⁻ or D⁻) are input into the CADSLAC code, which is a modification of the original electron trajectory code conceived at the Stanford Linear Acceleration Center, SLAC, by W B Hermannsfeldt [4]. A profile of the stripping losses (which depends on the source pressure) is calculated and also input into the code. CADSLAC then calculates the shape of the plasma boundary, which is determined by the V=0 potential. Then the beamlet trajectories through the pre-accelerator are calculated. The detailed result is an emittance diagram for <u>one</u> beamlet. The beamlet radius and the RMS divergence are used as input for the next stage, where 11 identical beamlets are simulated. In figure 2, the actual geometry of the pre-accelerator is given together with an example of the beamlet trajectories.

b) Simulate the post-acceleration stage

The beamlets calculated by CADSLAC are input into a simulation with the VectorFields OPERA-3D module SCALA [5]. This code takes the 11 beamlets with

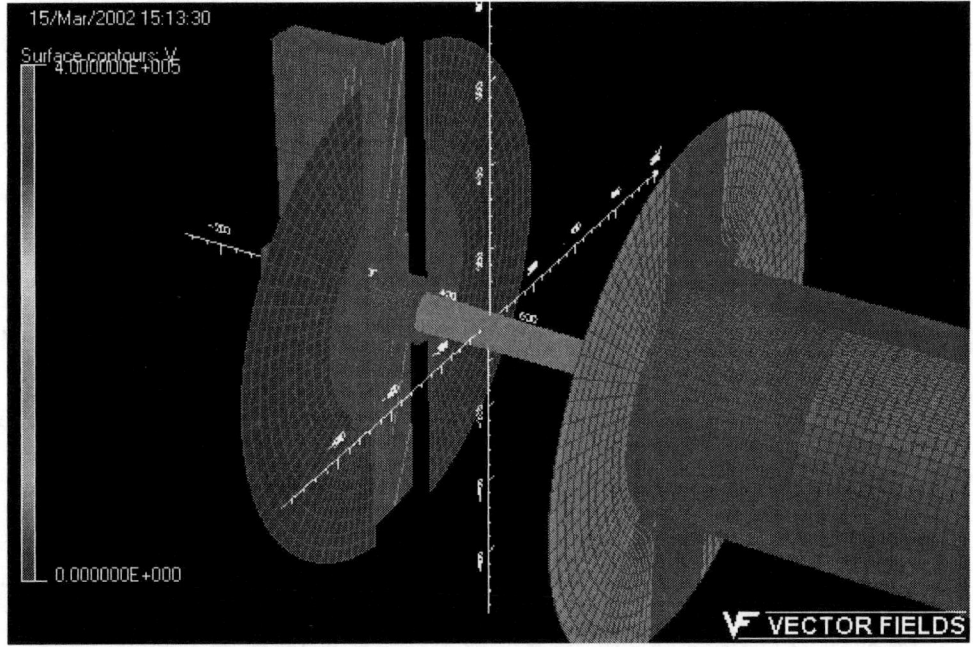

FIGURE 3: The 3-D model used to simulate the Cadarache SINGAP experiment.
A cylindrical kerb (inner diameter 106 mm) protrudes from the 43 keV pre-accelerator through the rectangular aperture in a grounded plate. It is arranged in this way in order to offer 'nice' equipotential surfaces to the beam. An 80 mm diameter aperture in the anode allows the beams to propagate to a target 2000 mm further downstream. Usually the simulation is started 105 mm inside the kerb and stopped 105 mm inside the anode.

the properties calculated by CADSLAC. Figure 3 shows the geometry. SCALA then calculates the propagation of the negative ion beams towards the target, taking the geometry in figure 3 into account. The result of the VectorFields calculation is an emittance diagram on which individual beamlets are usually recogniseable, even if they have merged in space.

Somewhere inside the kerb area, compensation of the beam space charge (by positive ions of the ionised background gas) can take place. Various assumptions are possible but SCALA cannot handle any of them. Therefore any simulation starts where space charge compensation inside the kerb is assumed to stop and the beam is allowed to drift between the pre-accelerator grid and this starting point. A simulation ends where space charge compensation inside the anode is assumed to begin. Due to the fields between kerb and anode, there can be no space charge compensation between the two. Where it begins and ends is a matter for assumption and so far the assumption has been one kerb diameter (105 mm) before the tip of the kerb.

For the same reason, the SCALA simulation is usually ended 105 mm beyond the entrance of the anode. Also, despite the 11 beamlets, symmetry along the horizontal and vertical axis is assumed and ¼ of the model is calculated (comprising 3 beamlets), i.e. the result is strictly only valid for 12 beamlets. The surplus beamlet is omitted in the next stage.

c) Beam transmission to the target.

The information from the VectorFields SCALA emittance diagram is used to define Gaussian beamlet profiles. These Gaussians are propagated through the anode to the target, using a purpose-written program called TRANSMIT. Because a code calculation is always idealised in some way (effects of source uniformity, ion temperature, magnetic field deflections, grid distortions, etc. are not considered), an extra divergence of 2 mrad is added to the beamlet parameters. Moreover, a random extra divergence between 0 and 1 mrad is added to the divergence of each beamlet. Also a random angle between -1 and +1 mrad is added to the starting angle of each beamlet. This is done to make the results more "realistic". TRANSMIT calculates a power density profile on the target and this can usually directly be compared with the measured data.

EXPERIMENTS UNDER LOW SPACE-CHARGE CONDITIONS

Effect of gap width

The gap in the Cadarache SINGAP assembly was changed during the winter of 2002 from 625 mm to 350 mm. An initial experiment was conducted on 22 March 2002. By comparing with data from the 2001 campaign, the effect of shortening the gap can be verified.

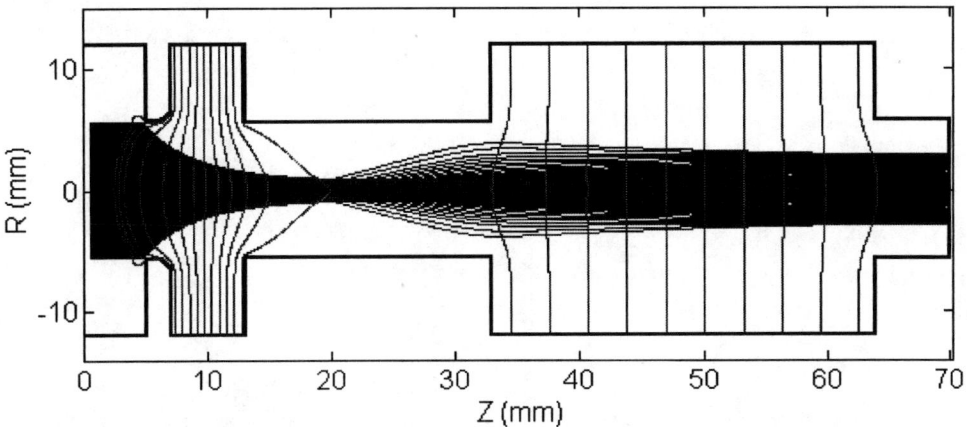

FIGURE 4: CADSLAC run for 45 kV pre-accelerator. The example is for shot 3938, where a 15 mA beam is extracted with V_{ext}=2.4 kV and pre-accelerated to V_{acc}=45 keV. (J_{ext}=1.35 mA/cm^2 D$^-$, 25% stripping losses result in J_{acc}=1 mA/cm^2). The calculated divergence of the resulting 45 keV beamlets is 12 mrad.

As previously, the pre-accelerator was simulated using CADSLAC. The simulation included stripping losses, which were calculated to be around 25%. The simulation for the pre-accelerator does not involve the post-accelerator voltage and/or gap. Thus, for identical pre-accelerator parameters (current density, source pressure, voltages) the beamlets produced must be the same and the effect of the post-accelerator voltage and gap can be studied separately. For the shots described here, the pre-accelerator parameters were constant and the beamlet is calculated to look like the one in figure 4.

Beamlet radius and divergence were input into the OPERA-3D SCALA module as described above. The emittance diagrams for the 575 keV beamlets are then obtained from OPERA-3D, see figure 5. The divergence of each individual beamlet is very small indeed, but their angular deflection is rather large: they cross-over strongly.

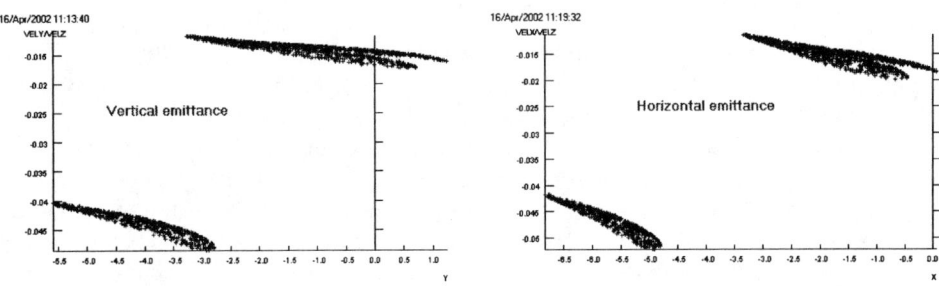

FIGURE 5: Emittance diagrams for the post-accelerated 1 mA, 575 keV beamlets from shot 3938.

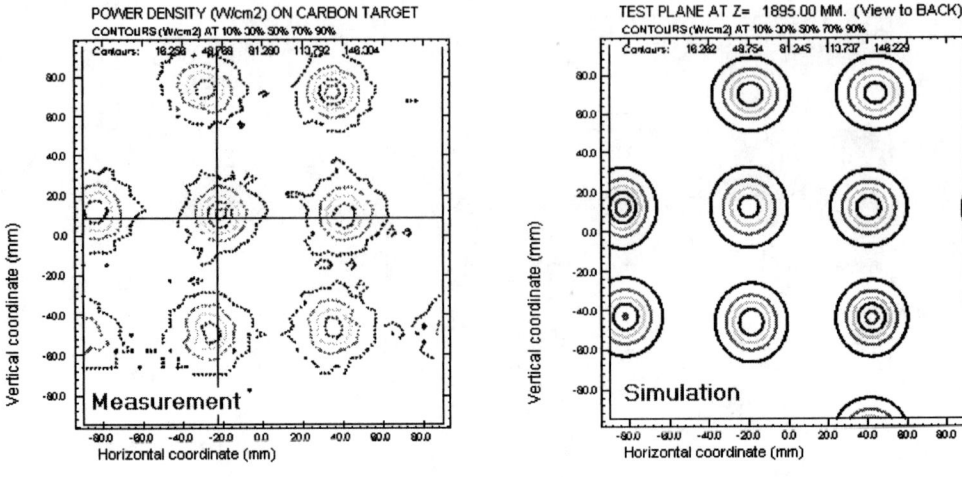

FIGURE 6: Measured and calculated power density contours for #3938.

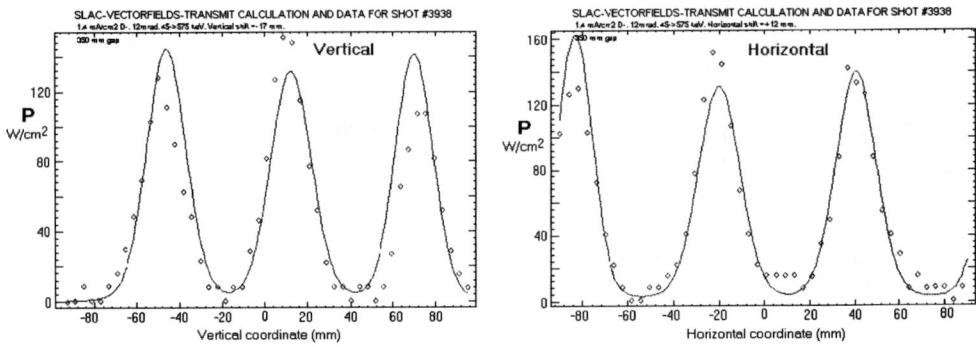

FIGURE 7: Measured (diamonds) and calculated (solid lines) power density profiles along the lines drawn in figure 6 for #3938.

Finally, the emittance diagrams were used to calculate the power density profiles on the carbon target, located 2 metres behind the anode aperture through which the beam enters. As the beamlets were calculated to be extremely narrorw, the heat diffusion inside the carbon target during the shot had to be calculated as well and taken into account. Measured and calculated power density profiles are in figures 6 and 7.

In figure 7 it can be seen that the agreement between experimental data and calculation is very good. The separation between the individual beamlets on the target is calculated to within 5%. Figure 8 shows two nearly identically shots, one from 2001 with the 625 mm gap, the other from 2002 with the 350 mm gap. It clearly demonstrates the effect of the gap on the electrostatic lens. Beamlets are much further apart with the short gap. Note that several of the 2002 beamlets miss the target altogether.

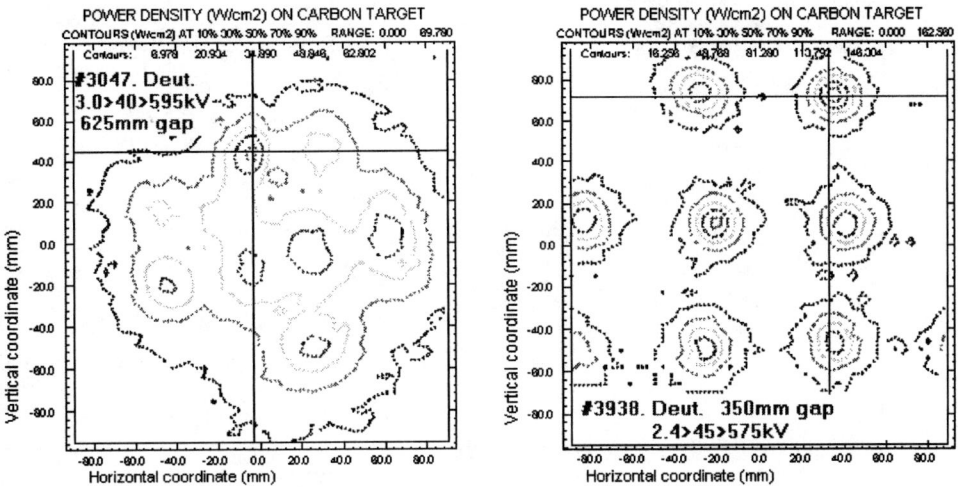

FIGURE 8: Power density profiles on the carbon target for nearly identical shots. Shot 3047 is from 2001 and has the 625 mm gap. Shot #3938 is from 2002 and has the 350 mm gap. The shorter gap clearly causes a stronger deflection of beamlets.

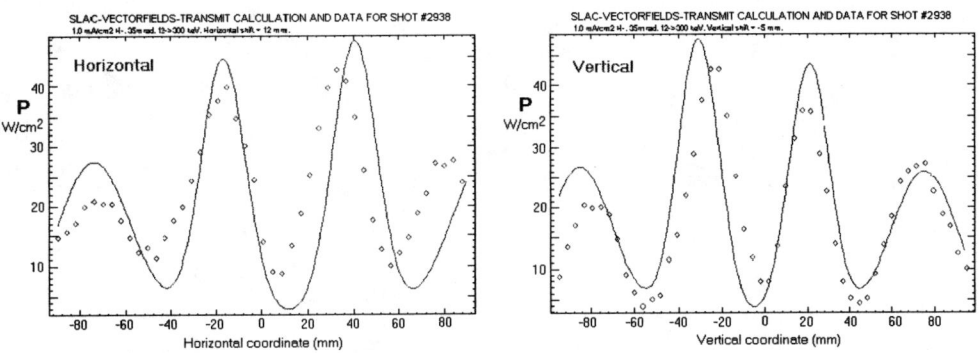

FIGURE 9: Experimental data and simulation for shot 2938 from the 2001 campaign, a similar low current density shot as the ones described above. With the 625 mm gap, the calculated distance between the beamlets is 14% too high, something found for all these shots.

Figure 9 shows a typical simulation from 2001, with the 625 mm gap. The calculated separation between the beamlets is 14% more than measured experimentally. Although only one example is shown here (in H$^-$ in stead of D$^-$), this discrepancy is observed throughout the campaign with the 625 mm gap.

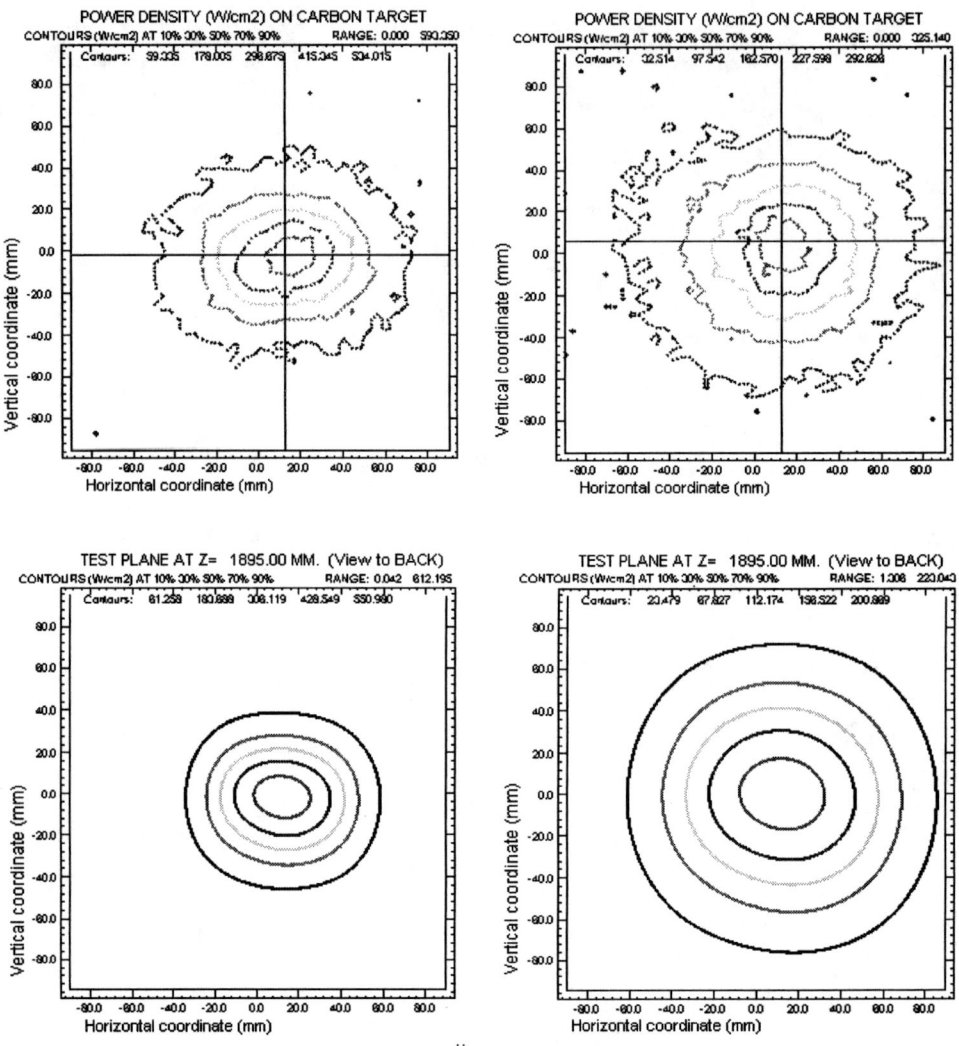

FIGURE 10: Measurements and simulations for two shots from the perveance scan. To the left shot 3668, with the calculated 10 mrad divergence pre-accelerated beamlets. To the right shot 3658 with 50 mrad pre-accelerated beamlets. The beam energy was 500 keV in both cases.

Perveance scan

On 10 December 2001, the extraction voltage V_2 was varied, V_3=43 kV, V_4=500 kV, $J(D^-)$=4 mA/cm^2. Simulation by CADSLAC resulted in beamlets leaving the pre-accelerator with divergences between 10 and 50 mrad, depending on V_2. The usual simulation procedure was followed, including the "realistic beams". The beamlets were not individually resolved. The SCALA calculations were done for 43 keV beamlet starting divergences of 10, 20, 30 and 50 mrad. It can be seen in figure 10 that, as expected, the 50 mrad spot is not 5 times as wide as the 10 mrad spot!

Gaussian fits were applied to both the experimental and simulated "blobs". The width of this data is plotted versus the extraction voltage V_2. The results are in figure 11. It can be seen that experimental data and calculations match quite well.

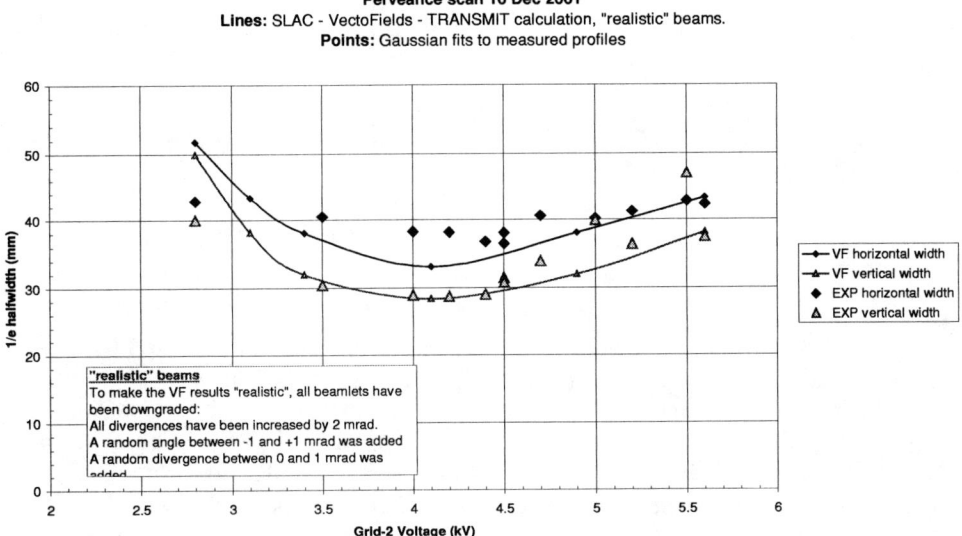

FIGURE 11: Perveance scan in extraction voltage. The widths are 1/e half widths on the target (either measured or simulated). The range in divergence of the pre-accelerated beamlets was 10 to 50 mrad.

MEDIUM SPACE-CHARGE EXPERIMENTS

So far, not many experiments could be done at medium space charge and none at high space charge. In this context, "medium" means a space charge of about 40% of the intended value on ITER and "high" means close to 100%.

The shot where the effect of space charge was most significant is 3332 (43→400 keV; gap=625 mm). The value $J_D^- = 5$ mA/cm^2 was obtained by measuring the heat dumped onto the graphite target, i.e. after ~20% stripping losses had occurred. Regarding space charge, this shot is equivalent to 8 mA/cm^2 D$^-$ being accelerated to 1 MeV, which is 40% of the ITER value 20 mA/cm^2.

CADSLAC calculates a divergence of 10 mrad for the 43 keV pre-accelerated beamlets. The calculation result is in figure 2. The 10 mrad divergence is used as input

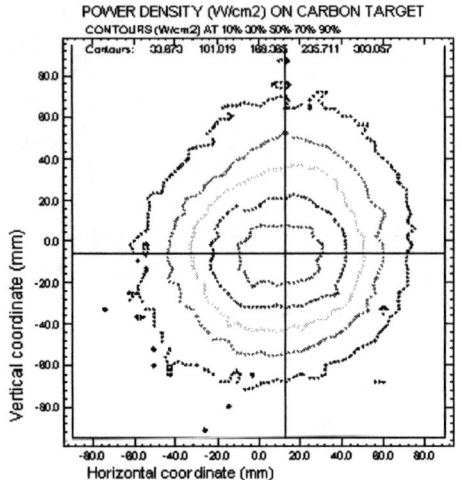 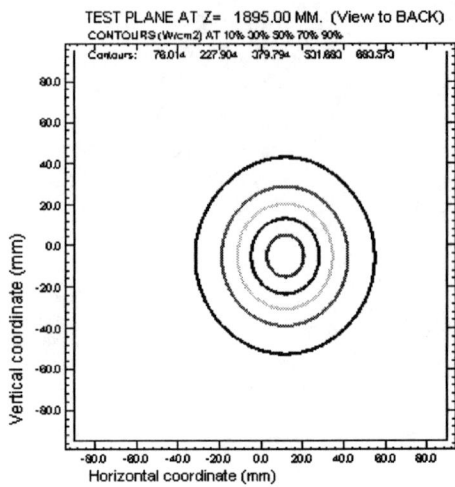

FIGURE 12: Measured and calculated power density profiles for #3332. The measured one is much wider than the calculated one.

to the OPERA-3D SCALA code [5]. The emittance diagrams calculated by SCALA were propagated to the target using the "realistic beams" assumption.

In figure 12, the simulation is compared with the experimental data. The simulation predicts a power density profile that is much too peaked. This is because in the simulation the space charge interaction between the beamlets prevents them from crossing over. The beamlets enter the anode structure quasi parallel and close together. Although the divergence of the individual post-accelerated beamlets is larger than in the low space-charge case, the resulting profile is (calculated to be) quite peaked.

In all the simulations so far, it has been assumed that space charge only acts inside the cylindrical kerb structure between the exit of the pre-accelerator grid and 105 mm before the tip of the kerb. Space charge compensation is assumed again beyond 105 mm inside the anode. In these regions it is assumed that the electrical fields are so weak that space charge compensation by the background gas can occur. Naturally, a few variations on this theme have been tried.

- A calculation without space charge resulted in narrow beamlets that are spread more apart. Although the result is a better fit to the data, it is not good.
- A calculation with space charge acting everywhere gave good results, see fig. 13.
- Re-calculating the low space-charge cases with space charge acting everywhere gave results conflicting with the data. Even in these low space charge cases, space charge can act strongly if narrow diameter beamlets leave the pre-accelerator.

Thus a <u>mild conflict</u> (50% difference) seems to be present in the sense that the same modelling assumption does not fit all the cases (low and medium space charge).

A possible explanation for this apparent conflict could be that there is not enough gas around to provide the positive ions necessary to compensate the beam. The gas pressure inside the Cadarache assembly is much lower than would be the case inside an ITER SINGAP system. It is conceivable that the medium space charge beams do

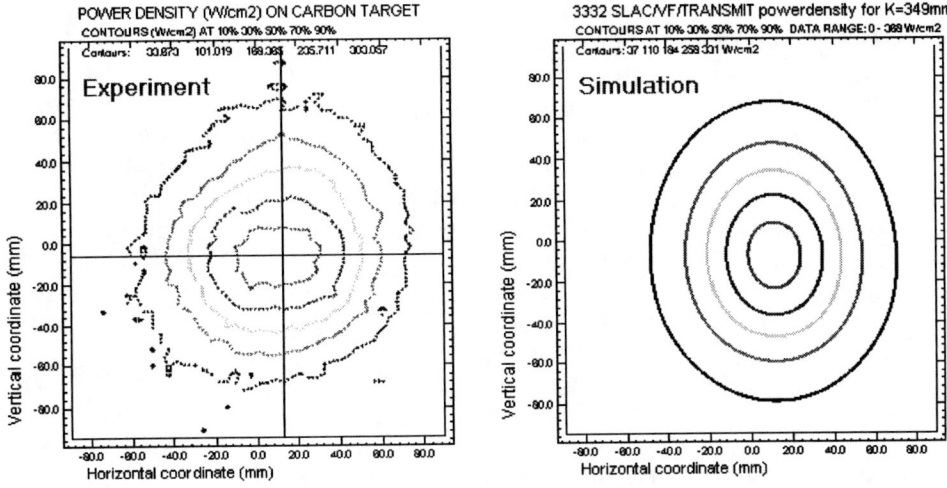

FIGURE 13: Measured and calculated power density profiles for #3332. In this simulation, space charge was allowed to act everywhere. This caused much more divergent post-accelerated beamlets leading to a profile that agrees much better with the data.

not get (sufficiently) compensated, whereas the low space-charge beams do. This assumption can be tested in experiments, which will be done at some point in the future.

CONCLUSIONS

The salient beam-optics conclusions from the 2001 campaign (with the 625 mm post-acceleration gap) and the first data from the 2002 campaign (with the 350 mm post-acceleration gap, data up to 1st June 2002) are:

- Electrostatic lensing effects generally hold. The difference between measured and simulated beamlet separation is 14% for the gap=625 mm data and 5% for the gap=350 mm data.
- The much larger beamlet separation for the gap=350 mm data was correctly reproduced by the simulations (figure 8).
- Beamlet widths were correctly simulated, provided that the "realistic beams" assumption was made.
- Perveance scans in extraction voltage V_2 were correctly simulated.
- The influence of different values for pre-acceleration voltage V_3 and post-acceleration voltage V_4 was correctly reproduced in simulations (although not all the different cases are shown in this paper).

The <u>modelling assumption</u> used was "space charge compensation in low field regions". It was implemented by enabling space charge compensation one kerb diameter (105 mm) away from the post-accelerator gap. This modelling assumption held well for low space-charge cases.

It did not hold very well for the few medium space-charge cases (J=5 mA/cm^2 D$^-$ at 400 keV). These cases were best simulated by enabling space charge <u>everywhere</u>. Therefore, a <u>mild conflict</u> exists between the modelling assumptions used. The apparent conflict could be caused by the generally very low gas pressure inside the SINGAP test stand and this will be tested in later experiments.

However, generally, the agreement between the data and simulations (with inconsistent assumptions about space charge compensation) is very good. The comparison now needs to be extended towards higher current density / space-charge conditions.

FUTURE WORK

The results presented here represent the beam-optics fruits from an experimental campaign started in September 2001. Technical limitations meant that neither the full ITER-relevant geometry nor current density could be used.

In the near future it is intended to run the negative ion source (the "Sourcette" [3]) at higher power, therefore higher D$^-$ currents and to continue the investigations at higher space charge with the 350 mm gap.

An upgraded version of the "Sourcette" is now designed. It should enable us to reach the full ITER relevant 20 mA/cm^2. Also a new pre-accelerator with new acceleration grids is being designed. The purpose of this is to remove the kerb structure that induces the strong electrostatic lenses.

In this way, a miniature 1 MeV SINGAP system that resembles the ITER SINGAP system closely should be available at Cadarache in 2003. This system should enable us to thoroughly test the ITER SINGAP beam optics and to assess whether a SINGAP system on ITER is viable.

REFERENCES

1. H.P.L. de Esch, R.S. Hemsworth and P. Massmann, "SINGAP, The European concept for negative ion acceleration in the ITER neutral injectors". *Rev. Sci. Instrum.* **73**(2002)1045.
2. J. Bucalossi, C. Desgranges, M. Fumelli, P. Massmann, and A. Simonin, *Rev. Sci. Instrum.* **70**(1999)1991.
3. A. Simonin, G. Delogu, C. Desgranges, M. Fumelli, "The DRIFT source : a negative ion source module for direct current multiampere ion beams", *Rev. Sci. Instrum.* **70**(1999)4542.
4. W.B. Hermannsfeld, "Electron Trajectory Program", *SLAC report* 226 (1979).
5. Vector Fields Ltd, 24 Bankside, Kidlington, Oxford OX5 1JE, UK. Tel: (+44)(0)1865 854999. http://www.vectorfields.co.uk

Negative Ion Production by fs, High-Intensity Laser Beam Interactions with Clusters

S. D. Moustaizis*[a#], Ph. Balcou*, J.-P. Chambaret*, D. Hulin*, G. Grillon*, J.-Ph. Rousseau*, and M. Schmidt[b]

Laboratoire d'Optique Appliquée, ENSTA- Ecole Polytechnique, CNRS Unité Mixte de Recherches 7639, F-91761 Palaiseau, France

[a] *Institute of Mater Structure and Laser Physics, Technical University of Crete, Kounoupidiana-Campus, 73100 Chania, Crete, Greece*

[b] *Service des Photons Atomes et Molécules, Centre d'Etudes de Saclay, F-91191-Gif-Sur-Yvette, France*

Abstract. We present experimental results concerning the observation and acceleration of negative and positive ions produced from high-intensity laser beam interactions with deuterated and hydrogenated clusters. The 30-fs, 10-Hz rep. rate, 30 TW Ti-Sa laser of LOA was focused on clusters produced from a pulsed gas nozzle. The novelty of our experimental work was the gas composition (CD_4 and CH_4), which was used for the cluster formation. Both kinds of ions, negative and positive, were observed with approximately equal energies. The maximum energy of both kinds of ions as well their energy distribution was measured using a Thomson parabola mass spectrometer. Negative and positive ions of H, D, and C were observed with a maximum energy up to 70 keV for a maximum laser beam intensity of 10^{18} W/cm^2. The mass spectrometer allowed us to correlate both the positive and negative ion energy spectrums with the laser beam intensity in order to study the process of cluster explosions, ion acceleration, and negative ion formation in the fs regime. We summarize the more important and new results as follows: (1) only single positive and negative ions were measured on the mass spectrometer; (2) the number of negative ions is equal to the number of positive ions; (3) similar energy spectrums for positive and negative ions were measured; (4) the positive and the negative ions are accelerated in the same radial direction, outwards from the interaction volume; and (5) similar energy spectrums for positive and negative ions at different laser energies (the laser intensity increases by a factor of 5) were measured. There are two possible "scenarios" concerning the process of negative ion formation. The first process is related with the formation of different ion species during the Coulomb explosion of the clusters. The second process for negative ion formation is the double electron attachment caused by collisions between the positive ions with the backing gas. A number of arguments confirm that the second process is more probable. The fact that from the mass spectrometer measurements the value for the more probable energy of the positive ions is the same as the energy at the maximum of the double electron attachment cross section leads to the second process of negative ion formation. For a complete description of our observation, however, we need an efficient physical process for negative ion formation, which will be the subject of our future work. The process of cluster explosion accelerates the D ions sufficiently to produce neutrons from DD nuclear fusion reactions. The efficiency of both the negative ions and the neutron alternative source will be discussed.

#e-mail: moustaiz@science.tuc.gr

Summary Comments

Joseph Sherman

Los Alamos National Laboratory, Los Alamos, NM 87545, USA

The Ninth International Symposium on the Production and Neutralization of Negative Ions and Beams was held in the Claude Bloch Auditorium on May 30–31, 2002, in Saclay, France. A good technical base for the continuation of negative hydrogen (H⁻) source development for the accelerator and magnetic confinement fusion activities was presented. Plenary oral sessions consisted of 21 contributions from 10 countries. R. Celiberto (Bari, Italy) addressed fundamental reaction processes in the gas phase, where electron-impact calculations are being collated with available data. O. Fukumasa (Yamaguchi University, Japan) discussed estimates for the transport of H⁻ in a source plasma to the extractor. A simulation model for a two-chamber system using a magnetic filter was described. Two experimental papers addressed the fundamental surface processes in H⁻ production. M. Nishiura (RIKEN, Japan) presented experimental work relating the reduction of the surface work function with increased H⁻ plasma density. This work used both rubidium and cesium alkali metals introduced into the plasma. Martha Bacal (Ecole Polytechnique, France) presented data on enhanced vibrationally excited H_2 molecules by vacuum ultraviolet (VUV) spectroscopy when a volume source is operated with tantalum filaments. The interpretation is a two-step process where recombinative desorption of hydrogen atoms from a tantalum wall film occurs to low vibrationally excited $H_2(v')$ molecules. This $H_2(v')$ population is then further excited by electron collisions leading to enhanced H⁻ production compared with H⁻ production by a tungsten filament source. A second presentation by O. Fukumasa used VUV spectroscopy in a two-chamber plasma system. The two chambers are separated by a grid used to inject electrons at controlled energies. VUV emission intensity is observed to increase with injected electron energy. Modeling of H⁻ extraction from the Cadarache SINGAP experiments was described by H. de Esch (CEA, Cadarache, Dutch). Commercially available codes SLAC and Vector Fields (Opera 3-D) are used with some success in modeling the SINGAP extraction and transport system. A theoretical treatment for extraction of volume-produced H⁻ ions based on an inverted plasma sheath was given by R. Becker (IAP, Frankfurt, Germany).

Also presented were several applications of H⁻ sources, the physics of which are based on the fundamental considerations discussed previously. The single example from the fusion community presented at this conference was given by Paul McNeely (Max-Planck Institut für Plasmaphysik, Garching, Germany—Canada). Plasma in a Joint European Torus (JET) bucket source is generated by an antennae external to the plasma, operating at 0.93 MHz with power up to 130 kW. Power is coupled through a ceramic, which is protected by an internal copper Faraday screen. A current density of 5–6 mA/cm^2 H⁻ current density has been extracted. Admixture of argon gas to the discharge enhances the H⁻ production. A second external antenna-driven plasma for H⁻ production was described by Y. S. Hwang (Seoul National University—Korea). Here the radio frequency (rf) power coupling is accomplished by a spiral antennae located on the back plane with a multi-cusp confinement system. Rf power is 13.56 MHz at 1–5 kW. Optimal H⁻ current of 3.3 mA is extracted at 30 kV by varying admixed argon gas and optimization of the magnetic filter field. Jens Peters (DESY, Germany) presented the new HERA H⁻ rf volume source, which couples low MHz rf power to the hydrogen plasma through an alumina ceramic. With the antenna located external to the plasma, a long lifetime (>25,000 hours) has been achieved in HERA accelerator operations. The source operates at a duty factor of 0.12% (8-Hz, 150-µs-long pulse), delivering 40 mA of H⁻ with approximately 40-kW of rf power. The plasma is in pure H_2 mode. Reliable pulsing is achieved through preionization of the H_2 feed gas before it is admitted into the main plasma chamber. The final rf-driven source was described by Roderich Keller (LBNL-SNS, United States). Plasma is produced by an insulator-coated rf antenna immersed in the active volume, and confinement is achieved by a multi-cusp magnet chamber. Cesium is introduced in a collar adjacent to the emission aperture. The cesium introduction enhances H⁻ production and reduces the extracted electron current. The source produces a 50-mA H⁻ beam at 65 keV, with the beam production efficiency being approximately 1 mA H⁻/kW rf power. This source has produced reliable beams of 35 mA and greater at the design duty factor of 6%. Reliable pulsing is achieved by keeping a low-density plasma present by a cw operation mode of 13.6-MHz power, which allows easy ignition of the plasma by the high-power, 2-MHz system. As much as a 50-mA beam has been transported through the front-end accelerator systems developed at LBNL for SNS. Klaus Volk (IAP, Frankfurt, Germany) described the Frankfurt filament-driven, cesiated H⁻ volume source. This source is equipped with a magnetic filter and an external steering magnet to dump electrons in transport. Up to 120 mA of H⁻ current has been measured. Although there has not been recent operation of this source, further development is possible through the European H⁻ support network. John Thomason (Rutherford Appleton Laboratories, England) described the ISIS (Rutherford-Appleton Laboratory's neutron spallation source) Penning surface plasma H⁻ source. This source produces 35 mA of H⁻ at 1% beam duty factor (200 µs, 50 Hz) with lifetimes of up to 50 days. This source is now injecting 36–37 mA of H⁻ into an rf quadrupole (RFQ), with 91% of transmissions being measured. Two advanced concepts for H⁻ production were presented at this conference. Raphael Gobin (CEA/Saclay, France) reported on the first H⁻ extracted from a 2.45-GHz microwave source. The order of 1-mA of H⁻ may have been found with 1 kW or less microwave power. The second advanced H⁻ generation technique was presented by S. Moustaizis (Technical University of Crete, Greece). This paper reported negative-ion generation from an ultrashort laser beam interacting with clusters. Negative ions are observed in a Thomson mass parabola.

Several papers were presented on the diagnostics of H⁻ beams. Douglas Moehs (Fermilab, United States) described a segmented Faraday cup that is being used to study the noise properties and beam space-charge neutralization effects for beams extracted from the Fermilab magnetron source. C. Gabor (IAP, Frankfurt, Germany) described a nondestructive photo detachment method for transverse beam emittance measurement. A. Jakob (IAP, Frankfurt, Germany) described the Frankfurt low-energy beam line and additional noninterceptive diagnostics for profile and beam potential measurements. M. Stockli (SNS, United States) described the SCUBEEX (self-consistent, unbiased elliptical exclusion analysis) method of emittance data analysis. This technique was applied to Berkeley SNS emittance data and showed that the Berkeley source meets the SNS emittance requirements. M. Stockli then presented for R. Welton (SNS, United States) analyses for a collection of H⁻ ion source emittance measurements made at major accelerator laboratories using the SCUBEEX technique. The majority of H⁻ beams from Penning, rf, or filament-driven volume sources and magnetrons gave similar emittance results when a consistent application of this program was made.

Conference attendees are grateful to CEA-Saclay for organizing and hosting the conference. The selected technical presentations were diverse and appropriate, the lunches were delicious, and the conference dinner was a delightful social experience. We are especially grateful to Raphael Gobin, Catherine Desailly, and Anne-Marie Gauriot for their efforts.

Symposium Program

Thursday, May 30, 2002

8:30-9:00		Registration
9:00-9:10		Welcome and Organization Details
9:10-9:35	R. Celiberto	Electron-Impact Cross Sections for Processes Involving Vibrational Excited Diatomic Hydrogen Molecules
9:35-10:00	O. Fukumasa	Modeling of Negative Ion Transport in Hydrogen Negative Ion Source - Estimation of Extracted H^- Current
		Coffee Break
10:20-10:45	M. Nishiura	Correlation of H^- Density and Work Function of a Plasma Electrode in H_2+ Alkali (Rb, Cs) Plasma
10:45-11:10	M. Bacal	Influence of Filament Material on VUV Emission in H^- Ion Source
11:10-11:35	H. De Esch	First Simulations of the Cadarache SINGAP Experiments
11:35-12:00	O. Fukumasa	Study of Isotope Effect on H^-/D^- Volume Production in Low Pressure H_2/D_2 Plasmas Using VUV Emission
		Lunch
13:50-14:15	P. Mac Neely	Extraction Studies on the Type VI RF Driven H^- Ion Source
14:15-14:40	R. Becker	An Inverted Plasma Sheath for the Simulation of the Extraction of Volume Produced H^-
14:40-15:05	D. Moehs	Studies on a Magnetron Beam
		Coffee Break
15:30-15:55	Y.S. Hwang	Characteristics and Optimization of a RF Negative Hydrogen Ion Source Using Transformer Coupled Plasma Sources
15:55-16:20	S. Moustaizis	Negative Ion Generation from Ultra-Short Laser Beam Interaction with Clusters
16:20-17:30	Open Discussion	Fundamental Processes: Wall versus Volume Production, Cs Effects, Wall Material, Magnetic Filter, Ta Collar, Gas Mixing

7:30 p.m.   **Conference Dinner**

Friday, May 31, 2002

8:45-9:10	J. Thomason	A Review of Recent H^- Ion Source Work at Rutherford Appleton Laboratory
9:10-9:35	J. Peters	The New HERA H^- RF Volume Source
9:35-10:00	R. Gobin	The CEA/Saclay 2.45 GHz Microwave Ion Source for H^- Ion Production
		Coffee Break
10:20-10:45	K. Volk	The Frankfurt H^- Source
10 45-11:10	C. Gabor	Status Report of the Frankfurt H^- Test LEBT Including a Non-destructive Emittance Measurement Device
11 10-11:35	A. Jakob	Diagnostic and Low Energy Beam Transport of H^- Ion Beams at Frankfurt
11:35-12:00	M. Stockli	Biasfree Analysis of Emittance Data Measured with a Single Current Amplifier
		Lunch
14:00-14:25	R. Keller	Design, Operational Experiences, and Beam Results Obtained with the SNS H- Ion Source and LEBT at Berkeley Lab
14:25-14:50	M. Stockli	Characterization of Emittance in High-Intensity H^- Ion Sources
		Coffee Break
15:15-16:30	**Open Discussion**	Beam Extraction and Transport Simulations, Electron Dumping
16:30-17:00	**Conclusion**	

List of Participants

Mme Bacal-Verney Marthe
Laboratoire LPTP,
Ecole Polytechnique,
91128 Palaiseau, France
Tel: (33) 1 69 33 32 52
Email: bacal@LPMI.Polytechnique.fr

Mr BEAUVAIS Pierre-Yves
CEA/Saclay
DAPNIA/SACM
91191 Gif-sur-Yvette Cédex, France
Tel: (33) 1 69 08 35 24
Email: beauvais@dapnia.cea.fr

BECKER Reinard
Institut für Angewandte Physik
Johann Wolfgang Goethe-Universität
Robert-Mayer Strasse 2-4
60054 Frankfurt/M – Germany
Tel: (49)-69-798-23488
Email: rbecker@physik.uni-frankfurt.de

BENMEZIANE Karim
CEA/Saclay
DAPNIA/SACM – Bt. 701
91191 Gif-sur-Yvette Cédex, France
Tel: (33) 1 69 08 68 20
Email: kbenmeziane@cea.fr

CELIBERTO Roberto
Politecnico di Bari
Dipartimento di Ingegneria Civile ed
Ambientale, (Sez. Chimica)
Via Orabona 4,
70125 Bari - ITALY
Tel: (39) 080 544 2104
Email: r.celiberto@cstar.poliba.it

CIAVOLA Giovanni
INFN-LNS
Via S. Sofia 44
CATANIA, Italy
Tel: (39) 095 542 262
Email: ciavola@lns.infn.it

COJOCARU Gabriel
TRIUMF
4004 Wesbrook Mall,
Vancouver, BC, Canada, V6T 2A3
Tel: (1) 604 222 61 63
Email: cojocaru@triumf.ca

DE ESCH Hubert
CEA Cadarache
DRFC
13 108 Saint Paul les Durance, France
Tel: (33) 4 42 25 46 69
Email: hplde@drfc.cad.cea.fr

FERDINAND Robin
CEA/Saclay
DSM/DAPNIA/SACM
91191 Gif-sur-Yvette Cédex, France
Tel: (33) 1 69 08 96 91
Email: ferdinand@dapnia.cea.fr

FUKUMASA Osamu
Department of Electrical & Electronic
Engineering
Yamaguchi University
Tokiwadai 2557, Ube 755, Japan
Tel: (81) 836 35 94 63
Email: fukumasa@plasma.eee.yamaguchi-u.ac.jp

GABOR Christoph
Institut für Angewandte Physik
Johann Wolfgang Goethe-Universität
Robert-Mayer Strasse 2-4
60054 Frankfurt/M – Germany
Tel: (49) 69-798 23475
Email: C.Gabor@iap.uni-frankfurt.de

GEBEL Ralf
Forschungszentrum Juelich GmbH
Institut für Kernphysik - COSY
Leo-Brandt-Str.
D-52428 Jülich, Germany
Tel: (49) 02461 61 3097
Email: r.gebel@fz-juelich.de

GIRARD Alain
CEA Grenoble
Rue des Martyres
38 000 Grenoble, France
Tel: (33) 4 38 78 43 65
Email: girard@drfmc.ceng.cea.fr

GOBIN Raphael
CEA/Saclay
DSM/DAPNIA/SACM
91191 Gif-sur-Yvette Cédex, France
Tel: (33) 1 69 08 27 64
Email: gobin@dapnia.cea.fr

HWANG Yong-Seok
Seoul National University
San 56-1, Shillim-dong
Kwanak-ku
Seoul 151-742, Korea
Tel: (82)-2-880-6276
Email: yhwang@snu.ac.kr

IVANOV Andrey
Laboratoire LPTP,
Ecole Polytechnique,
91128 Palaiseau, France
Tel: (33) 1 69 33 32 52
Email: ivanov@lptp.polytechnique.fr

JAKOB Ansgar
Institut für Angewandte Physik
Johann Wolfgang Goethe-Universität
Robert-Mayer Strasse 2-4
60054 Frankfurt/M – Germany
Tel: (49) 69 798 23475
Email: A.Jakob@iap.uni-frankfurt.de

KELLER Roderich
LBNL, SNS Front-End Systems
Berkeley Lab,
1 Cyclotron Road, MS 71-259
Berkeley, CA 94720, USA
Tel: (1) 510.486.5223
Email: r_keller@lbl.gov

KRAUS Werner
Max-Planck-Institut für Plasmaphysik,
Boltzmannstrasse 2
85748 Garching, Germany
Tel: (49) 89 3299 2243
Email: werner.kraus@ipp.mpg.de

KUCHLER Detlef
Particle Production Group
PS Division
CERN
CH-1211 Geneva 23, Switzerland
Tel: 41 22 767 6691
Email: Detlef.Kuchler@cern.ch

McNEELY Paul
Max-Planck-Institut für Plasmaphysik
Boltzmannstrasse 2
85748 Garching, Germany
Tel: (49) 89 3299 1339
Email: p.mcneely@ipp.mpg.de

MOEHS Douglas
FERMILAB
Fermi National Accelerator Laboratory,
P.O. Box 500,
Batavia, IL 60510 USA
Tel: (1)-630 840 4490
Email: moehs@fnal.gov

MOUSTAIZIS Stavros
Institute of Mater Structure and Laser Physics,
Technical University of Crete
Kounoupidiana-Campus,
73100 Chania, Crete, Greece
Tel: 30 821 0 28450
Email: moustaiz@science.tuc.gr

NISHIURA Masaki
RIKEN
Institute of Physical and Chemical Research,
2-1 Hirosawa, Wako, Saitama 351-0198
Japan
Tel: (81) 484 674 232
Email: mn@postman.riken.go.jp

OLIVO Miguel
TRIUMF
4004 Wesbrook Mall,
Vancouver, BC, Canada, V6T 2A3
Tel: (1) 604 222 74 85
Email: olivo@triumf.ca

PETERS Jens
DESY
Notkestr. 85
22607 Hambourg, Germany
Tel: (49) 40 8998 2495
Email: jens.peters@desy.de

ROBICHE Jérôme
Plasma Research Laboratory
Dublin City University
Glasnevin, Dublin 9, Ireland
Tel:
Email: Jerome.Robiche@dcu.ie

SCHMIDT Charles
FERMILAB
Fermi National Accelerator Laboratory,
P.O. Box 500,
Batavia, IL 60510 USA
Tel: (1) 630 840 4414 (4441)
Email: cschmidt@fnal.gov

SHERMAN Joe
Los Alamos National Lab,
LANSCE-2. MS H838
Los Alamos, NM 87545, USA
Tel: (1) 505-667-3511
Email: jsherman@lanl.gov

STOCKLI Martin P.
SNS - Spallation Neutron Source
Accelerator Systems Division
115 Union Valley Rd, MS6471
Oak Ridge, TN 37830, USA
Tel: (1) 865-241-8817
Email: stockli@sns.gov

THOMASON John
ISIS,
Rutherford Appleton Laboratory
Didcot,
Oxfordshire,
OX11 0QX, United Kingdom.
Tel: (44) 1235 446050
Email: J.W.G.Thomason@rl.ac.uk

VOLK Klaus
Institut für Angewandte Physik
Johann Wolfgang Goethe-Universität
Robert-Mayer Strasse 2-4
60054 Frankfurt/M, Germany
Tel: 069-7982-3395
Email: k.volk@iap.uni-frankfurt.de

Author Index

A

Alessi, J., 160

B

Bacal, M., 13
Balcou, P., 197
Bandyopadhyay, M., 90
Becker, R., 82
Benmeziane, K., 177
Boilson, D., 184

C

Celiberto, R., 3
Chambaret, J.-P., 197

D

de Esch, H. P. L., 184
Delferrière, O., 177

F

Ferdinand, R., 177
Franzen, P., 90
Fujioka, T., 75
Fukuchi, T., 75
Fukumasa, O., 28, 75

G

Gabor, C., 121, 128
Glass-Maujean, M., 13
Gobin, R., 177
Grillon, G., 197

H

Harrault, F., 177
Heinemann, B., 90
Hemsworth, R. S., 184
Hong, I. S., 61
Hu, C., 90
Hulin, D., 197
Hwang, Y. S., 61

I

Ivanov Jr., A. A., 13

J

Jakob, A., 121, 128
Jung, H. D., 61

K

Keller, R., 47, 135, 160
Klein, H., 67, 121, 128
Klomp, A., 121
Kraus, W., 90

L

Laricchiuta, A., 3
Letchford, A. P., 135

M

Maaser, A., 67
Massmann, P., 184
McNeely, P., 90
Meusel, O., 121, 128
Moehs, D. P., 115
Mori, S., 28
Moustaizis, S. D., 197

N

Nishiura, M., 13, 21

P

Park, Y. S., 61
Peters, J., 42
Pozimski, J., 121, 128

R

Ratzinger, U., 67, 121, 128
Riedl, R., 90
Rousseau, J.-P., 197

S

Santić, F., 121
Sasao, M., 13, 21
Schäfer, J., 121
Schmidt, M., 197
Sherman, J. D., 160, 177, 198
Speth, E., 90

Stockli, M. P., 47, 135, 160
Svensson, L., 184

T

Takeiri, Y., 28
Tauchi, Y., 28
Thomae, R. W., 47, 135, 160
Thomason, J. W. G., 37, 135, 160

V

Volk, K., 67

W

Wada, M., 13, 21
Welton, R. F., 47, 135, 160
Wilhelm, R., 90

Y

Yabuki, Y., 28